*Design of Structural Steelwork*

Consulting Editor: Ray Hulse, Coventry University

Other titles of related interest:

**Basic Solid Mechanics**  D.W.A. Rees

**Civil Engineering Construction – Design and Management**  D.R. Warren

**Civil Engineering Contract Administration and Control, Second Edition**  I.H. Seeley

**Civil Engineering Materials, Fourth Edition**  N. Jackson and R. Dhir (Editors)

**Civil Engineering Quantities, Fifth Edition**  I.H. Seeley

**Design of Structural Timber**  W.M.C. McKenzie

**Finite Elements – A Gentle Introduction**  David Henwood and Javier Bonet

**Highway Traffic Analysis and Design, Third Edition**  R.J. Salter and N. Hounsell

**Plastic Analysis of Steel and Concrete Structures, Second Edition**  S.J.J. Moy

**Soil Mechanics – Principles and Practice**  G. Barnes

**Reinforced Concrete Design, Fourth Edition**  W.H. Mosley and J.H. Bungey

**Reinforced Concrete Design to Eurocode 2**  W.H. Mosley, R.Hulse and J.H. Bungey

**Structural Masonry, Second Edition**  Arnold W. Hendry

**Structural Mechanics**  J. Cain and R. Hulse

**Surveying for Engineers, Third Edition**  J. Uren and W.F. Price

**Understanding Hydraulics**  L. Hamill

**Understanding Structures, Second Edition**  Derek Seward

# Design of Structural Steelwork
## to BS 5950 and C-EC3

W.M.C. McKenzie BSc, PhD, CPhys, MInstP, CEng
*Lecturer, Napier University, Edinburgh*

First published 1998 by
MACMILLAN PRESS LTD
Houndmills, Basingstoke, Hampshire RG21 6XS
and London
Companies and representatives throughout the world

ISBN 0–333–71579–9 ✓ SRB

A catalogue record for this book is available from the British Library.

This book is printed on paper suitable for recycling and made from
fully managed and sustained forest sources.

10   9   8   7   6   5   4   3   2
07   06   05   04   03   02   01   00

Printed and bound in Great Britain by
Antony Rowe Ltd, Chippenham, Wiltshire

*To my parents Jeannie and Robert*

# Contents

# Preface

Most engineers/engineering technicians develop their initial understanding of structural design by undertaking a recognised, structured undergraduate course of study. The material covered in such courses requires the student to refer to and use relevant British Standard Codes of Practice. Many practising engineers/technicians do not become involved in design immediately following graduation and require reference material with which to update and refresh their knowledge of design.

This text has been prepared for both types of reader. It is suitable for students undertaking degree or diploma level studies in architecture, civil, structural or building engineering.

The aim of the text is to develop an understanding of Limit State Design as applied to structural steelwork. It is assumed that readers will have access to and use the relevant British Standard Codes of Practice, i.e. BS 5950:Part 1, BS 6399:Parts 1 and 2 and the Concise Eurocode C-EC3.

The British Standards Institution publishes a selection of extracts from various codes entitled *Extracts from British Standards for students of structural design*; students using this document will be undertaking a course of study which is less extensive than the range covered by the present text and they may choose to ignore the additional more detailed material and clause references that it includes.

The use of the codes, particularly BS 5950:Part 1 is explained and demonstrated in numerous worked example and illustrations. The text is both extensive and comprehensive; it includes explanations of structural analysis and load distribution techniques where considered appropriate. Student readers are strongly advised to read, use and familiarise themselves with the terminology adopted in design codes by frequent references to relevant clauses throughout the text. Structural design cannot be efficiently carried out without reference to design codes.

The complexity and nature of modern design codes necessitates reference to additional material for full explanations of the rationale behind their development; appropriate sources are provided in the bibliography.

A selection of design examples from Chapters 2, 5 and 6 are presented in a format typical of that used in a design office in order to encourage students to adopt a methodical and rational approach in preparing structural calculations. The solutions for these examples are given at the end of their respective chapters.

W.M.C. M<sup>c</sup>Kenzie

# Acknowledgements

The author and the publishers are grateful to the following organisations for permission to reproduce copyright material in this book.

Extracts from BS 5950: Part 1: 1990 are reproduced with the permission of BSI. Complete editions of the standards can be obtained by post from BSI Customer Services, 389 Chiswick High Road, London W4 4AL.

Extracts of tables from
*Steelwork design guide to BS 5950: Part 1: Volume 1*
*C-EC3 – Concise Eurocode for the design of steel buildings in the United Kingdom*
have been reproduced by kind permission of the Director, The Steel Construction Institute. Complete copies can be obtained from The Steel Construction Institute, Silwood Park, Ascot, Berks SL5 7QN.

# 1. Structural Steelwork

## 1.1 Introduction

Modern steelmaking processes employing blast furnaces to produce pig-iron from iron-ore, coke and limestone are refinements of those methods used by early ironworkers. Despite smelting techniques being used for several thousand years it is only relatively recently (i.e. the 1800s), with William Kelly's 'pneumatic' steelmaking process in America, and shortly after, Sir Henry Bessemer's converter in England, and the Siemens-Martin open-hearth process, that methods of mass production of steel have evolved.

Essentially, the raw materials are heated to temperatures in the region of 1500°C. During this heating process the coke releases carbon monoxide, which when combined with the iron oxides in the ore, produces metallic iron. Impurities (e.g. silica which combines with the limestone to form calcium silicate) form a slag which floats to the top of the molten metal and is removed leaving 'pig-iron' behind. Further refinements of the pig-iron, in which it is combined with scrap steel and iron ore are carried out to remove excess carbon and other impurities such as silica, phosphorus, manganese and sulphur. The resulting molten steel can be poured into cast-iron moulds to form ingots or continuously cast into slabs which are subsequently cut to required lengths.

The precise method of refinement used depends on the type of steel required. There are five main categories:

(i) **carbon steels** containing varying amounts of carbon, manganese, silica and copper;

(ii) **alloy steels** as in (i) and in addition containing varying amounts of vanadium, molybdenum or other elements;

(iii) **high-strength low-alloy** steels as in (i) and in addition containing small amounts of expensive alloying elements;

(iv) **stainless steels** as in (i) and in addition containing chromium, nickel and other alloying elements;

(v) **tool steels** as in (i) and in addition containing tungsten, molybdenum and other alloying elements.

Each type of steel exhibits different characteristics such as strength, ductility, hardness, and corrosion resistance. The most widely used type of structural steelwork is carbon steel, although stainless steel is frequently used for architectural reasons and has been used on some of the world's most famous structures, e.g. the Chrysler Building in New York and the Canary Wharf Tower in London. In addition to aesthetic considerations, stainless steel significantly reduces the maintenance requirements and costs during the lifetime of a structure.

The physical properties of various types of steel and steel alloys depend primarily on the percentage and distribution of carbon in the iron; in particular the relative proportions of ferrite, pearlite and cementite, all of which contain carbon. Heat treatment of steel is carried out to control the amount, size, shape and distribution of these substances and to develop other compounds such as austenite and martensite. This treatment hardens and strengthens the steel but in addition induces additional residual strains. Following heat treatment, tempering (annealing) is normally carried out to relieve these strains. This consists of reheating to a lower temperature which results in a reduction in hardening and strength but produces an increase in ductility and toughness.

The first major structural use of modern steel in Britain was in the construction of the Forth Railway Bridge in Scotland. This impressive elegant structure designed by Baker and Fowler makes use of tubular and angular sections riveted and forged together to form the appropriate structural elements. The development of hot-rolling, welding and bolting techniques during the twentieth century has resulted in a vast range of available cross-section and fabrication options which enable designers to produce a limitless variety of structures reflecting the lightness, stiffness and strength of modern steel.

Steel is in many respects an ideal structural material which exhibits a number of useful characteristics such as:

♦ structural behaviour up to initial yield which is linearly elastic,
♦ deformations which are directly proportional to applied loads, and are fully recovered on removal of the loading,
♦ the yield strength in tension is approximately equal to the crushing strength in compression,
♦ isotropic behaviour,
♦ long-term deformation (creep) at normal temperatures is not a problem.

In addition, the plastic deformation which steel structures exhibit after initial yield gives a margin of safety and early warning that the structure is stressed beyond the elastic limit, and that loads should be removed and the structure should be strengthened.

As a result of these properties, structures in steelwork can be analysed using a variety of classic elastic analysis techniques, elasto-plastic and plastic methods. The deformations and stresses predicted using such techniques have good correlation with experimentally measured values.

There are some aspects of the behaviour of steel structures which require more rigorous analysis than is possible using the principles of statics and linear elastic techniques. When structural steel sections, which may be hot-rolled or cold formed into thin sections, are subject to stress concentrations at locations such as support points or re-entrant corners, local yielding may occur with a subsequent redistribution of loading through the structure. A result of this is that these structural elements are susceptible to failure by buckling under compression, lateral torsional buckling (bowing laterally along the member length) or bearing (crinkling) locally under concentrated loads.

Steel is a material which offers a designer wide scope to create original and unique structural forms and systems. There are, however, several structural forms which have evolved to fulfil clearly defined requirements, and which have a widely adopted range of standardised structural sections and connection details. In addition to the requirements of economy and speed of construction on site, a client may have specific reasons for choosing a steel-framed structural system; for example a large column-free internal space, ease in accommodating future expansion in both the horizontal and vertical directions, demountability, ease of transport and handling on site.

Technical developments in the production, fabrication, finishing and on-site construction of steel have had a significant impact on the economics of steel design. Typical of such developments are:

♦ Steel-rolling mills which are highly automated, as are the fabrication workshops, thereby improving product quality and reducing the cost of finished steel sections for delivery to site,

♦ Specialist companies exist which design, fabricate and in some cases, deliver and erect, widely used special structural members and systems,

♦ A variety of special alloy steels for specific functional or aesthetic applications have been developed. Many structural steel sections are for example available in several grades of stainless steel containing a percentage of chromium and/or nickel for use in hostile corrosive environments. *Weathering* steel contains a small percentage of copper which permits an outer layer of corrosion to develop on the exposed surfaces which inhibits further deterioration within the section,

♦ Fully automated, computer-controlled cutting and welding of steel sections, particularly rolled hollow sections, has led to the development of prefabricated trusses and space-frame modules. The extensive use of High-Strength Friction-Grip bolts and to a lesser extent site welding, is now widely adopted to achieve efficient and economic rigid connections.

The two most widely used applications of structural steelwork in buildings are:

(i) Single-storey industrial/commercial warehouse facilities, which may be braced or rigid-jointed as shown in Figure1.1, and

(ii) Multi-storey braced frames in which the lateral wind loads are resisted by steel bracing elements or reinforced concrete shear walls as shown in Figure 1.2, or alternatively rigid-jointed frames in which lateral stability is provided by moment connections between the columns and floor beams.

In braced frames the lateral loading is assumed to be transmitted by the vertical elements, (i.e. diagonal bracing and/or reinforced concrete shear walls), while the vertical load is transmitted by simple shear connections between the columns and beams. This enables a simple static distribution of floor and roof loadings to supporting beams, as indicated in Section 1.2.3.

In rigid-jointed frames, sway forces induced by the lateral loading are transmitted throughout the frame by providing full moment and shear connections at an appropriate number of beam to column connections to ensure continuity of the structure. This technique requires a more complex analysis procedure to determine the design bending moments, shear forces and axial loads.

Both systems have advantages and disadvantages; the most appropriate one for any given structure will depend on a number of factors, such as the need for adaptability, speed and ease of fabrication, contract programme constraints and fire-proofing requirements.

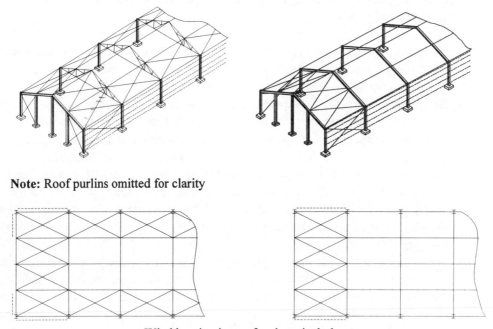

**Note:** Roof purlins omitted for clarity

Wind bracing in  roof and vertical planes

Single-storey braced frame                     Single-storey rigid jointed frame

**Figure 1.1**

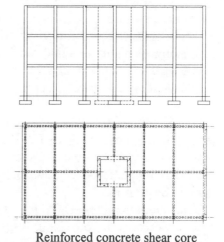

Multi-storey braced frames with K-bracing          Reinforced concrete shear core

**Figure 1.2**

Some types of building have a structural form and proportions which lend themselves to a particular system, but there is inevitably some overlap between structural systems and applications.

The adoption of a particular structural form or system may be the result of economic or architectural style as much as structural and functional necessity. In the 1970s significant investment in modernising and renewing urban light rail systems resulted in a number of bright, attractive rail stations with space deck roofs, such as those built for the Newcastle Metro and the Glasgow Underground railways. In the 1980s there was a major drive to develop London's Docklands as a new commercial and financial centre, resulting in the widespread construction of 'fast-track' multi-storey steel frames with composite floor construction. In the 1990s, extensive reconstruction of football stadia has taken place following the Taylor Report (Ref:21) on crowd safety at football grounds. Again steel-framed construction has been adopted to take advantage of the speed of erection.

## 1.2  The Design Process

The design of a structure is initiated by the acceptance of a client's brief defining specific requirements. The brief will generally indicate the use which is to be made of the structure(s), e.g. an office block with specified usable floor areas, a commercial development utilizing a limited plot of land, or a transportation development scheme comprising a variety of structures such as bridges, terminal buildings and associated facilities.

Preliminary designs and scheme drawings are prepared and considered by the design team, which may include structural engineers, architects, local authority representives, the potential user and the client.

After evaluation of the possible alternatives, including an estimate of budget costs and approval by the client, final detailed design and preparation of working drawings is carried out. Most contracts go through a tender stage in which a number of contractors are asked to submit a tender price (bid) to carry out the construction work; this sometimes involves design and build arrangements. There are numerous variations in the precise details for which bids are made. Site supervision of the successful contractor is normally carried out by a representative of the consultant engaged by the client.

### 1.2.1  Aims of Design

The aim of design is to produce structures which

- ◆ satisfy the clients' needs during their intended life,
- ◆ sustain, with adequate margins of safety, all the loads and deformations expected during their use,
- ◆ are resistant to disproportionate damage caused by events such as explosions, impacts or accidents,

♦ are serviceable and durable throughout their design life,
♦ are economic to construct and maintain,
♦ enhance, where possible, their environment or minimize their intrusive effects.

These aims can be achieved by careful consideration being paid to the selection of materials, adopting recognised good practice in design and detailing techniques and using control procedures for all stages of the design and construction process.

### 1.2.2 Structural Loading

All structures are subjected to loading from various sources. The main categories of loading are: dead, imposed and wind loads. In some circumstances there may be other loading types which should be considered such as settlement, fatigue, temperature effects, dynamic loading, impact effects, e.g. when designing bridge decks, crane-gantry girders or maritime structures. In the majority of cases design considering combinations of dead, imposed and wind loads is the most appropriate.

The definition of dead and imposed loading is given in BS 6399:Part 1:1984, while wind loading is defined in BS 6399:Part 2:1995. Part 2 is a technical revision of, and supersedes, CP3:Chapter V:Part 2:1972. Both wind codes are currently in use, but the intention is to withdraw CP3:Chapter V: Part:2 1972.

### 1.2.2.1  Dead Loads:  BS 6399:Part 1: 1984      (Clause 3.0)

Dead loads are loads which are due to the effects of gravity, i.e. the self-weight of all permanent construction such as beams, columns, floors, walls, roof and finishes etc.
If the position of permanent partition walls is known their weight can be assessed and included in the dead load. In speculative developments, internal partitions are regarded as imposed loading.

### 1.2.2.2  Imposed Loads:  BS 6399:Part 1: 1984      (Clauses 4.0 and 5.0)

Imposed loads are loads which are due to variable effects such as the movement of people, furniture, equipment and traffic. The values adopted are based on observation and measurement and are inherently less accurate than the assessment of dead loads.
*Clause 4.0* and *Tables 5* to *12* define the magnitude of uniformly distributed and concentrated point loads which are recommended for the design of floors, ceilings and their supporting elements. Loadings are considered in the following categories:
Residential:
Type 1    Self-contained dwelling units
Type 2    Apartment houses, boarding houses, guest houses, hostels, lodging houses, residential clubs and communal areas in blocks of flats.

Type 3   Hotels and motels
Institutional and educational
Public assembly
Offices
Retail
Industrial
Storage
Vehicular

Most floor systems are capable of lateral distribution of loading and the recommended concentrated load need not be considered. In situations where lateral distribution is not possible the effects of the concentrated loads should be considered, with the load applied at locations which will induce the most adverse effect, e.g. maximum bending moment, shear and deflection. In addition, local effects such as crushing and punching should be considered where appropriate.

In multi-storey structures it is very unlikely that all floors will be required to carry the full imposed load at the same time. Statistically, it is acceptable to reduce the total floor loads carried by a supporting member by varying amounts depending on the number of floors or floor area carried. This is reflected in *Clause 5.0* and *Tables 2* and *3* of BS6399:Part 1, in which a percentage reduction in the total distributed imposed floor loads is recommended when designing columns, piers, walls, beams and foundations. It should be noted that the loadings given in *Table 3* do not include the imposed load carried by the roof.

### 1.2.2.3  *Imposed Roof Loads  BS 6399:Part 3: 1995    (Draft version)*

Imposed loading caused by snow is included in the values given in this part of the code which relates to imposed roof loads. Flat roofs, sloping roofs and curved roofs are also considered. This part of BS 6399 is a **draft version** for public comment and should **not** be used as a British Standard.

### 1.2.2.4  *Wind Loads:  BS 6399:Part 2: 1995*

Environmental loading such as wind loading is clearly variable and its source is beyond human control. In most structures the dynamic effects of wind loading are small and static methods of analysis are adopted. The nature of such loading dictates that a statistical approach is the most appropriate in order to quantify the magnitudes and directions of the related design loads. The main features which influence the wind loading imposed on a structure are:

♦ geographical location   – London, Edinburgh, Inverness, Chester ...
♦ physical location      – city centre, small town, open country, ...

- ◆ topography            – exposed hill top, escarpment, valley floor, ...
- ◆ building dimensions    – height above mean sea level,
- ◆ building shape         – square, rectangular, cruciform, irregular, ...
- ◆ roof pitch             – shallow, steep, mono-pitch, duo-pitch, multi-bay ...
- ◆ wind speed and direction
- ◆ wind gust peak factor.

Tabulated procedures enable these features to be evaluated and hence produce a system of equivalent static forces which can be used in the analysis and design of the structure.

### 1.2.3  Structural Analysis and Load Distribution

The application of the load types discussed in Section 1.2.2 to structural frames results in axial loads, shear forces, bending moments and deformations being induced in the floor/roof slabs, beams, columns and other structural elements which comprise a structure. The primary objective of structural analysis is to determine the distribution of internal moments and forces throughout a structure such that they are in equilibrium with the applied design loads. There are a number of mathematical models, some manual and others computer based, which can be used to idealise structural behaviour. These methods include: two- and three-dimensional elastic behaviour, elastic behaviour considering a redistribution of moments, plastic and non-linear behaviour. Detailed explanations of these techniques can be found in the numerous structural analysis text books which are available; they are not explained in this text.

In braced structures where floor slabs and beams are considered to be simply supported, vertical loads give rise to three basic types of beam loading condition:

(i)    uniformly distributed line loads,
(ii)   triangular and trapezoidal loads,
(iii)  concentrated point loads.

These load types are illustrated in Examples 1.1 to 1.4.

### 1.3  Example 1.1  Load distribution – one-way spanning slabs

Consider the floor plan shown in Figure 1.3 where two one-way spanning slabs are supported on three beams A-B, C-D and E-F. Both slabs are assumed to be carrying a uniformly distributed design load of 5 kN/m$^2$.

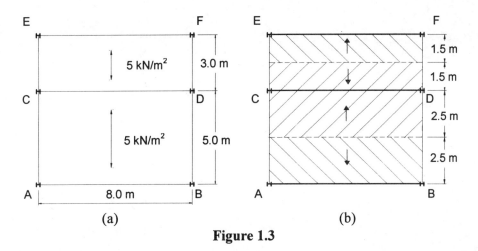

**Figure 1.3**

Both slabs have continuous contact with the top flanges of their supporting beams and span in the directions indicated. The floor area supported by each beam is indicated in Figure 1.3 (b).

Beam AB
  Total load  =  (floor area supported × magnitude of distributed load/m$^2$)
            =  (2.5 × 8.0) × (5.0)  =  100 kN

Beam CD
  Total load  =  (4.0 × 8.0) × (5.0)  =  160 kN

Beam EF
  Total load  =  (1.5 × 8.0) × (5.0)  =  60 kN

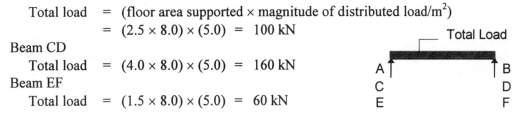

Check: the total load on both slabs = (8.0 × 8.0 × 5.0)  =  320 kN

**Note:** Beams AC, CE, BD and DF are assumed to be supporting only their self-weight in addition to loading which may be applied directly such as brickwork walls.

## 1.4  Example 1.2  Two-way spanning slabs

Consider the same floor plan as in Example 1.1 but in this case the floor slabs are two-way spanning as shown in Figure 1.4.

Since both slabs are two-way spanning, their loads are distributed to supporting beams on all four sides assuming a 45° dispersion as indicated in Figure 1.4(b).

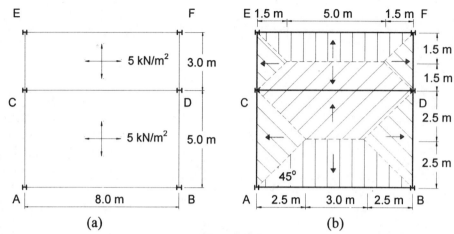

**Figure 1.4**

**Beam AB**

Load due to slab ABCD $= \left(\dfrac{8.0+3.0}{2} \times 2.5\right) \times (5.0)$

$= 68.75 \text{ kN}$

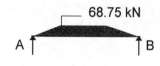

**Beam EF**

Load due to slab CDEF $= \left(\dfrac{8.0+5.0}{2} \times 1.5\right) \times (5.0)$

$= 48.75 \text{ kN}$

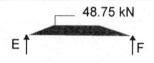

**Beams AC and BD**

Load due to slab ABCD $= \left(\dfrac{5.0}{2} \times 2.5\right) \times (5.0)$

$= 31.25 \text{ kN}$

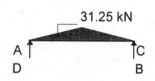

**Beams CE and DF**

Load due to slab CDEF $= \left(\dfrac{3.0}{2} \times 1.5\right) \times (5.0)$

$= 11.25 \text{ kN}$

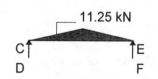

The loading on beam CD can be considered to be the addition of two separate loads, i.e.
Load due to slab ABCD     = 68.75 kN     (as for beam AB)
Load due to slab CDEF     = 48.75 kN     (as for beam EF)

Check: the total load on all beams $\;=\;$ $2(68.75 + 48.75 + 31.25 + 11.25)$
$\qquad\qquad\qquad\qquad\qquad\qquad\quad = \;$ 320 kN

## 1.5  Example 1.3  Secondary beams

Consider the same floor plan as in Example 1.1 with the addition of a secondary beam spanning between beams AB and CD as shown in Figure 1.5. The load carried by this new beam imposes a concentrated load at the mid-span position of both beams AB and CD.

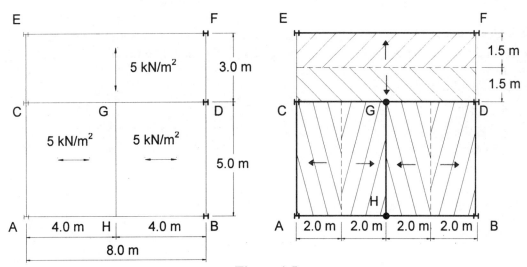

**Figure 1.5**

Beam EF
$\qquad$ Total load $\quad = \;$ $(1.5 \times 8.0) \times (5.0)$
$\qquad\qquad\qquad\qquad = \;$ 60 kN

Beam GH
$\qquad$ Total load $\quad = \;$ $(4.0 \times 5.0) \times (5.0)$
$\qquad\qquad\qquad\qquad = \;$ 100 kN

Beams AC and BD
$\qquad$ Total load $\quad = \;$ $(2.0 \times 5.0) \times (5.0)$
$\qquad\qquad\qquad\qquad = \;$ 50 kN

Beam AB
$\qquad$ Total load $\quad = \;$ End reaction from beam GH
$\qquad\qquad\qquad\qquad = \;$ 50 kN

Beam CD
The loading on beam CD can be considered to be the addition of two separate loads, i.e.

Load due to beam GH   =  50 kN   (as for beam AB)
Load due to slab CDEF =  60 kN   (as for beam EF)

## 1.6  Example 1.4  Combined one-way, two-way slabs and beams

Considering the floor plan shown in Figure 1.6, with the one-way and two-way spanning slabs indicated, determine the type and magnitude of the loading on each of the supporting beams.

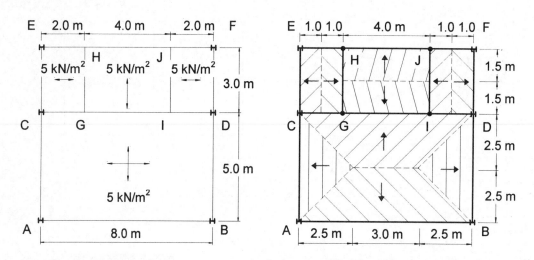

**Figure 1.6**

The loads on beams AB, AC and BD are the same as in Example 1.2.

Beams CE, DF, GH and IJ

$$\text{Total load} = (3.0 \times 1.0) \times (5.0)$$
$$= 15 \text{ kN}$$

Beam EF

End reaction from beam GH = 7.5 kN
End reaction from beam IJ  = 7.5 kN

Load from slab GHIJ     = $(4.0 \times 1.5) \times (5.0)$
                         = 30 kN

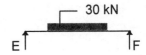

Total loads on beam EF due to beams GH, IJ and slab GHIJ

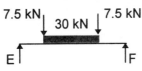

Beam CD

    End reaction from beam GH = 7.5 kN
    End reaction from beam IJ  = 7.5 kN

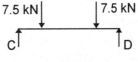

Load from slab GHIJ = $(4.0 \times 1.5) \times (5.0)$ = 30 kN

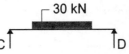

Trapezoidal load from slab ABCD is the same as in Example 1.2

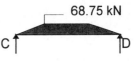

Total loads on beam CD due to beams GH, IJ and slab GHIJ and ABCD

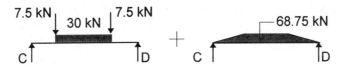

## 1.7  Limit State Design  BS 5950:Part 1: 1990

The synthesis of a particular design solution evolves from the interaction of a wide range of considerations to ensure that a structure does not become unsatisfactory in use during its intended lifetime. The concepts of limit state design examine conditions which may be considered to induce failure of a structure; such conditions are known as *limit states*. In BS 5950:Part 1: 1990 *Structural use of steelwork in building* two types of limit state are classified, they are:

♦ Ultimate limit states

♦ Serviceability limit states.

### 1.7.1 Ultimate Limit State

This is a consideration of the strength and/or stability of a structure and its component parts at the onset of collapse and includes effects such as yielding, rupture, buckling, local and overall instability, fatigue and brittle fracture.

### 1.7.2 Serviceability Limit State

This is a consideration of the response of a structure and its component parts while in service and includes effects such as deflection, vibration, corrosion or any behaviour which renders the structure unsuitable for its intended purpose when subjected to service loads. The general principles of limit state design as applied to the structural steelwork design of buildings are given in *Section two* of BS 5950:Part 1.

### 1.7.3 Partial Safety Factors

When checking the suitability of a structure or structural element with respect to a particular limit state it is necessary to introduce appropriate safety factors. Although not stated explicitly in BS 5950:Part 1: 1990, it is generally accepted that material strengths and applied loading which have **not** been factored are referred to as **characteristic values** and those which have been multiplied by a safety factor as **design** values.

The inherent variability in material strengths and applied loads (as discussed in section 1.2), and in assessing their magnitudes, dictates that different safety factors should be used to reflect different circumstances. The introduction of *Partial Safety Factors* applied to different types of loading conditions and material types is a feature of limit state design.

In BS 5950 these factors have been simplified to allow for a number of effects such as mathematical approximations in analysis and design techniques and tolerances in material manufacture, structural fabrication and erection methods. Material design strengths are given in *Table 6* of BS 5950 and include the appropriate partial safety factors, while the design loadings are obtained by multiplying the characteristic loads by the appropriate partial safety factors ($\gamma_f$) given in *Table 2*.

A detailed explanation of the rationale adopted by BS 5950 in determining the recommended partial safety factors is given in *Appendix A* of the code.

### 1.7.4 Application of Partial Safety Factors

The application of the partial safety factors to applied loading, in addition to BS6399:Parts 1 and 2 referred to in section 1.2.2, is illustrated in Example 1.5.

### 1.8 Example 1.5 Brewery part floor plan load distribution

A part-plan of a brewery building is shown in Figure 1.7. Using the data provided and the relevant BS codes, determine the design shear force and bending moment for beams CD, EF and AB.

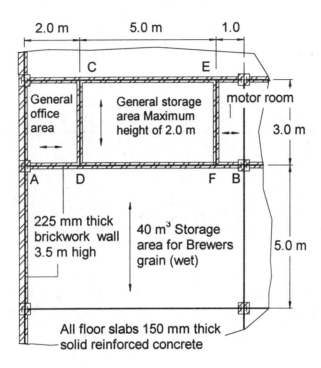

**Figure 1.7**

**Solution:**

| | | |
|---|---|---|
| Unit weight of wet brewers grain | = | $5.5$ kN/m$^3$ |

(this information is available from the client)

Unit weight of concrete = $24.0$ kN/m$^3$

Self-weight of 150 mm thick R.C. slab = $0.15 \times 24.0$ = $3.6$ kN/m$^2$

Unit weight of 100 mm thick brickwork = $2.2$ kN/m$^2$ of surface area

Self-weight of 225 mm thick brickwork = $\dfrac{2.2 \times 225}{100}$ = $4.95$ kN/m$^2$

**BS 5950:Part 1      Table 2**

Partial safety factor for dead loads      = 1.4

Partial safety factor for imposed loads      = 1.6

In all three cases the critical load combination is the combined dead and imposed loading. The design loads can therefore be determined by adding the factored dead and imposed load, i.e.

Design load = (1.4 × dead load) + (1.6 × imposed load)

**BS 6399:Part 1: 1984**

**Table 10:**   Imposed load in motor room   = $7.5$ kN/m$^2$

|               | General storage                         | $= 2.4 \times 2.0$ | $= 4.8 \ kN/m^2$ |
|---------------|-----------------------------------------|--------------------|------------------|
| **Table 8**   | Imposed load in general office          |                    | $= 2.5 \ kN/m^2$ |
|               | Brewers grain (wet)                     | $= 5.5 \times 40$  | $= 220 \ kN$     |

**Beam CD**

| Design dead load due to self-weight of slab   | $= 1.4 \times 3.6 \times 3.0$  | $= 15.12 \ kN$ |
|-----------------------------------------------|--------------------------------|----------------|
| Design dead load due to brick-work wall       | $= 1.4 \times 4.95 \times 10.5 = 72.77 \ kN$ |  |
| Design imposed load due to general office     | $= 1.6 \times 2.5 \times 3.0$  | $= 12.0 \ kN$  |

$$\text{Total design load} = ( 15.12 + 72.77 + 12.0 )$$
$$= 100 \ kN$$

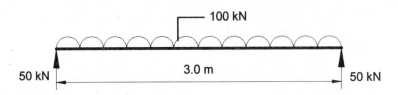

100 kN

3.0 m

50 kN                              50 kN

Design shear force       $= 50 \ kN$

Design bending moment $= \dfrac{WL}{8} = \dfrac{(100 \times 3.0)}{8} = 37.5 \ kNm$

**Beam EF**

| Design dead load due to self-weight of slab   | $= 1.4 \times 3.6 \times 1.5$  | $= 7.56 \ kN$ |
|-----------------------------------------------|--------------------------------|---------------|
| Design dead load due to brickwork wall        | $= 1.4 \times 4.95 \times 10.5 = 72.77 \ kN$ |  |
| Design imposed load due to motor room         | $= 1.6 \times 7.5 \times 1.5$  | $= 18.0 \ kN$ |

$$\text{Total design load} = (7.56 + 72.77 + 18.0)$$
$$= 98.3 \ kN$$

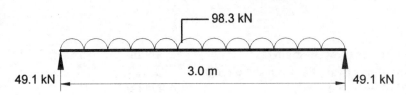

98.3 kN

3.0 m

49.1 kN                              49.1 kN

Design shear force       $= 49.1 \ kN$

Design bending moment $= \dfrac{WL}{8} = \dfrac{(98.3 \times 3.0)}{8} = 36.9 \ kNm$

**Beam AB**

Design dead load due to self-weight of slab in grain storage area

|                                               | $= 1.4 \times 3.6 \times 2.5 \times 8.0$  | $= 100.8 \ kN$ |
|-----------------------------------------------|-------------------------------------------|----------------|
| Design dead load due to brickwork wall        | $= 1.4 \times 4.95 \times 8.0 \times 3.5$ | $= 194.0 \ kN$ |
| Design imposed load due to wet grain          | $= 1.6 \times 110.0$                      | $= 176.0 \ kN$ |

$$\text{Total design load} = (100.8 + 194.0 + 176.0)$$
$$= 470.8 \ kN$$

Design dead load due to self-weight of slab in general storage area

$$= 1.4 \times 3.6 \times 1.5 \times 5.0 \qquad = 37.8 \text{ kN}$$

Design imposed load due to general storage

$$= 1.6 \times 4.8 \times 1.5 \times 5.0 \qquad = 57.6 \text{ kN}$$

Total design load $= (37.8 + 57.6) \qquad = 95.4 \text{ kN}$

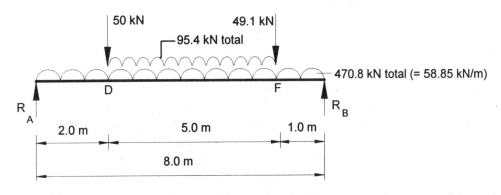

$$R_B = \frac{\left[(50 \times 2.0) + (470.8 \times 4.0) + (95.4 \times 4.5) + (49.1 \times 7.0)\right]}{8} = 344.5 \text{ kN}$$

$$R_A = (665.3 - 344.5) = 320.8 \text{ kN}$$

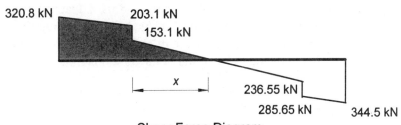

Shear Force Diagram

**Note:**

The value of the UDL between D and F $= \dfrac{95.4}{5} + \dfrac{470.8}{8} = 77.93 \text{ kN/m}$

$x = \dfrac{153.1}{77.93} = 1.96 \text{ m}$ this is the position of the maximum bending moment

Design Shear Force $= 344.5 \text{ kN}$

Design bending moment is equal to the shaded area in the shear force diagram.

Design Bending Moment $= \left[\dfrac{320.8 + 203.1}{2}\right] 2 + \left[\dfrac{1.96 \times 153.1}{2}\right] = 673.94 \text{ kNm}$

## 1.9 Application of Wind Loads

Two methods of analysis for determining the equivalent static wind loads on structures are given in BS 6399:Part 2: 1995, they are:

♦ the standard method
♦ the directional method

The standard method is a simplified procedure and is intended to be used by designers undertaking hand-based calculations. This method gives conservative results for a wide range of the most commonly used structures.

The directional method, while assessing wind loads more accurately, is more complex and is intended for use with computational analysis; it is not considered in this text.

Structures which are susceptible to dynamic excitation by virtue of their structural properties, e.g. mass, stiffness, natural response frequencies or structural form such as slender suspended bridge decks or long span cable stayed roofs will generally require more complex mathematical analysis techniques and/or wind tunnel testing. The standard method is illustrated in Examples 1.6 to 1.8.

## 1.10 Example 1.6 Storage hopper

A closed top storage hopper as shown in Figure 1.8 is situated in an industrial development near Edinburgh and adjacent to the sea. Assuming the altitude of the location to be 5.0 m above mean sea level, determine the overall horizontal wind loading on the structure, while considering the wind to be acting in the direction indicated.

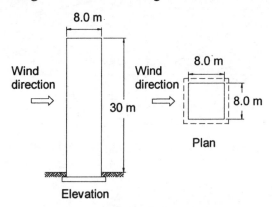

**Figure 1.8**

**Solution:**
*Clause 2.2.3.2*
Since the crosswind breadth (8.0 m) is less than the height (30.0 m) a reduction in lateral loading is permitted.
*Figure 11(c)* of BS 6399:Part 2     $H = 30$ m,        $B = 8.0$ m $\therefore H > 2B$
**Note: this applies to pressures only and does not apply to suctions**

Consider the building surface to be divided into a number of parts A, B, C and D

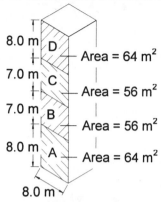

8.0 m D — Area = 64 m²

7.0 m C — Area = 56 m²

7.0 m B — Area = 56 m²

8.0 m A — Area = 64 m²

8.0 m

**Figure 1.9**

*Clause 2.1.3.6*
The overall load on the building $\quad P = 0.85(\Sigma P_{front} - \Sigma P_{rear})(1 + C_r)$
where:

$\Sigma P_{front}$ is the sum of the horizontal components of surface load on the windward facing wall

$\Sigma P_{rear}$ is the sum of the horizontal components of surface load on the leeward facing wall

$C_r$ is a dynamic augmentation factor

*Clause 2.1.3.5*
Net load on an area of surface $\quad P = pA$
where:

$p$ is the net pressure across the surface
$A$ is the area of surface being considered

*Clause 2.2.1    Figure 6*
For Edinburgh the basic wind speed $\quad V_b = 23.5$ m/sec

*Clause 2.2.2.2.2    Figure 7*
Assuming that the topography of the site is not considered significant
$$\text{altitude factor} \quad S_a = 1 + 0.001\Delta_s$$
$$= 1 + (0.001 \times 5)$$
$$S_a = 1.005$$

*Clause 2.2.2*
$$\text{site wind speed} \quad V_s = V_b \times S_a \times S_d \times S_s \times S_p$$
In many cases the direction factor ($S_d$), seasonal factor ($S_s$) and probability factor ($S_p$) can be considered to be equal to 1.0 (see *Clauses 2.2.2.3 to 2.2.2.5*).
$$V_s = (23.5 \times 1.005)$$
$$= 23.62 \text{ m/sec}$$

*Design of Structural Steelwork*

*Clause 2.2.3*

$$\text{effective wind speed} \quad V_e = V_s \times S_b$$

where:

$S_b$  is the terrain and building factor obtained from *Clause 2.2.3.3* and *Table 4*

*Clause 2.2.2.3  Table 4*

The effective wind speed for each of the areas A, B, C and D can be determined assuming an effective height $H_e$ equal to the reference height $H_r$ to the top of each area.

*Table 4*

| | | |
|---|---|---|
| Area A | $H_e = 8.0$ m | $S_b = 1.73$ |
| Area B | $H_e = 15.0$ m | $S_b = 1.85$ |
| Area C | $H_e = 22.0$ m | $S_b = 1.91$ |
| Area D | $H_e = 30.0$ m | $S_b = 1.96$ |

Effective wind speeds

| | |
|---|---|
| Area A | $V_e = 23.62 \times 1.73 = 40.9$ m/sec |
| Area B | $V_e = 23.62 \times 1.85 = 43.7$ m/sec |
| Area C | $V_e = 23.62 \times 1.91 = 45.1$ m/sec |
| Area D | $V_e = 23.62 \times 1.96 = 46.3$ m/sec |

*Clause 2.1.2.1  Table2*

Dynamic wind pressure  $q_s = 0.613V_e^2$

| | |
|---|---|
| Area A | $q_s = 0.613 \times 40.92 = 1.03$ kN/ m$^2$ |
| Area B | $q_s = 0.613 \times 43.70 = 1.17$ kN/m$^2$ |
| Area C | $q_s = 0.613 \times 45.10 = 1.25$ kN/m$^2$ |
| Area D | $q_s = 0.613 \times 46.30 = 1.31$ kN/m$^2$ |

*Clause 2.1.3.1*

External surface pressure    $p_e = q_s C_{pe} C_a$

where:

$C_{pe}$  is the external pressure coefficients (*Clause 2.4*)
$C_a$   size effect factor (*Clause 2.1.3.4*)

**Note:** In this problem it is not necessary to consider internal pressure coefficients since the overall horizontal loading is being considered. Internal pressure coefficients are considered in Example 1.7.

*Clause 2.1.3.4*    *Figures 4 and 5*

Diagonal dimension      $a = \sqrt{30^2 + 8^2} = 31.05$ m

From *Figure 4* use *line A* in the graph to determine $C_a \approx 0.89$

*Clause 2.4*      *Figure 12* and *Table 5*

*Table 5*        $\dfrac{D}{H} = \dfrac{8}{30} = 0.27 < 1.0$

               $C_{pe}$  windward face  $= +0.8$

               $C_{pe}$  leeward face   $= -0.3$

**Note:**    +ve  indicates pressure on a surface

           −ve  indicates suction on a surface

In this problem the wind loading on the side faces and roof are not being considered.

External surface pressure          $p_{front} = q_s \times 0.8 \times 0.89 = +0.712q_s$

                                      $p_{rear} = q_s \times 0.3 \times 0.89 = -0.267q_s$

| | | |
|---|---|---|
| Area A | $p_{front} = +0.712 \times 1.03$ | $= +0.73 \text{ kN/m}^2$ |
| | $p_{rear} = -0.267 \times 1.31$ | $= -0.35 \text{ kN/m}^2$ |
| Area B | $p_{front} = +0.712 \times 1.17$ | $= +0.83 \text{ kN/m}^2$ |
| | $p_{rear} = -0.267 \times 1.31$ | $= -0.35 \text{ kN/m}^2$ |
| Area C | $p_{front} = +0.712 \times 1.25$ | $= +0.89 \text{ kN/m}^2$ |
| | $p_{rear} = -0.267 \times 1.31$ | $= -0.35 \text{ kN/m}^2$ |
| Area D | $p_{front} = +0.712 \times 1.31$ | $= +0.93 \text{ kN/m}^2$ |
| | $p_{rear} = -0.267 \times 1.31$ | $= -0.35 \text{ kN/m}^2$ |

*Clause 2.1.3.5*

Net load on building surface area   $= p \times$ loaded area

| | | | | |
|---|---|---|---|---|
| Area A | $P_{front} = 0.73 \times 64 = +47.72 \text{ kN}$ | $P_{rear} = -0.35 \times 64 = -22.40 \text{ kN}$ |
| Area B | $P_{front} = 0.83 \times 56 = +46.48 \text{ kN}$ | $P_{rear} = -0.35 \times 56 = -19.60 \text{ kN}$ |
| Area C | $P_{front} = 0.89 \times 56 = +49.84 \text{ kN}$ | $P_{rear} = -0.35 \times 56 = -19.60 \text{ kN}$ |
| Area A | $P_{front} = 0.93 \times 64 = +59.52 \text{ kN}$ | $P_{rear} = -0.35 \times 64 = -22.40 \text{ kN}$ |

*Clause 2.1.3.6*

The overall horizontal load on the building   $P = 0.85(\Sigma P_{front} - \Sigma P_{rear})(1 + C_r)$

*Clause 1.6.1*    *Table 1*  *Figure 3*

*Table 1*      Building type factor             $K_b = 1.0$

*Figure 3*     Dynamic augmentation factor    $C_r \approx 0.04$

$\Sigma P_{front} = +(47.72 + 46.48 + 49.84 + 59.52) = +203.56 \text{ kN}$

$\Sigma P_{rear} = -(22.40 + 19.60 + 19.60 + 22.40) = -84.00 \text{ kN}$

Overall horizontal load   $P = 0.85(203.56 - (-84.0)) \times (1 + 0.04)$

                           $P = 254.2 \text{ kN}$

## 1.11  Example 1.7  Industrial warehouse

An industrial warehouse is to be designed comprising a series of three-pinned pitched roof portal frames as shown in Figure 1.10. Using the data provided determine the wind loading on the structure.

**Design Data:**

| | |
|---|---|
| Location | open country near Preston |
| Altitude of site | 20.0 m above mean sea level |
| Closest distance to sea | 2.0 km |
| Overall length of building | 24.0 m |
| Centres of frames | 4.0 m |

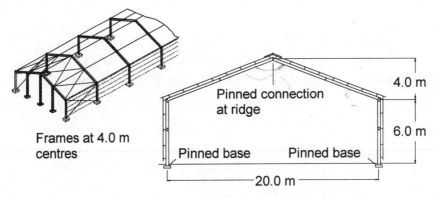

Frames at 4.0 m centres

Pinned connection at ridge

Pinned base      Pinned base

4.0 m

6.0 m

20.0 m

Typical internal frame

**Figure 1.10**

**Solution:**

*Clause 2.2.3.2*

*Figure 11(a)*    $H = 10$ m        $B = 24$ m        $H < B$

∴ consider building as one part

*Clause 2.2.1    Figure 6*

Location    Preston    Basic wind speed    $V_b = 23.0$ m/sec

*Clause 2.2.2.2.2    Figure 7*

Assuming that the topography is not considered significant

$\Delta_s = 20.0$ m        Altitude factor        $S_a = 1 + (0.001 \times 20.0)$

$= 1.02$

*Clause 2.2.2*

Site wind speed $V_s = V_b \times S_a \times S_d \times S_s \times S_p$

Assuming        $S_d = S_s = S_p = 1.0$

$V_s = 23.0 \times 1.02 = 23.46$ m/sec

*Clause 2.2.3   Table 4*
  Effective wind speed      $V_e = V_s \times S_b$

*Clause 2.2.3.3*          $H_e = H_r = $ actual height  $= 10.0$ m (see *Clause 1.7.3.2*)
*Table 4*                closest distance to sea   $= 2$ km      $\therefore S_b = 1.78$
                $V_e = 23.46 \times 1.78$   $= 41.76$ m/sec

*Clause 2.1.2.1*
  Dynamic wind pressure                    $q_s = 0.613 V_e^2$
                                $= (0.613 \times 41.76^2)/10^3 = 1.07$ kN/m$^2$

*Clause 2.1.3.1*
  External surface pressures   $p_e = q_s C_{pe} C_a$

*Clause 2.1.3.2*
  Internal surface pressures   $p_i = q_s C_{pi} C_a$

## External Pressure Coefficients
Consider the wind acting on the longitudinal face of the building

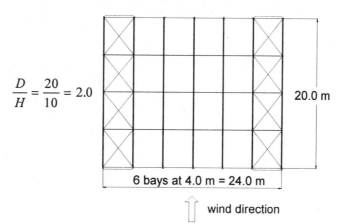

$$\frac{D}{H} = \frac{20}{10} = 2.0$$

20.0 m

6 bays at 4.0 m = 24.0 m

$\Uparrow$ wind direction

**Figure 1.11**

*Clause 2.4  Table 5*

The values of $C_{pe}$ in *Table 5* can be interpolated in the range $1 < \dfrac{D}{H} < 4$

$\uparrow 0.23$

leeward face       $C_{pe} = -0.23$
windward face     $C_{pe} = +0.73$

$\uparrow 0.73$        **Figure 1.12**

In the case of the gable faces, different values of $C_{pe}$ should be used depending on the gap between adjacent buildings; in this example assume that the building is isolated.

*Clause 2.4.1.3      Figure 12*
The surfaces on which the wind is blowing are considered as separate zones which are
defined in *Figure 12* and depend on a variable '*b*',
where:

$$b \leq \text{crosswind breadth} \quad = \quad 24.0 \text{ m}$$
$$\leq 2 \times \text{height} \quad\quad = \quad 2 \times 10.0 \quad = \quad 20.0 \text{ m}$$

Use the smaller value of *b*   i.e.          *b*   =   20.0 m

*Figure 12(b)*    *D* = *b* ∴ consider the gable surface as two zones A and B

The width of zone A is equal to   $0.2 \times b$   =   $0.2 \times 20$     =   4.0 m
The width of zone B is equal to                    =   $(20.0 - 4.0)$   =   16.0 m

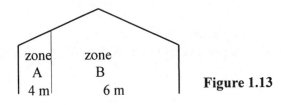

zone A    4 m
zone B    6 m

**Figure 1.13**

gable faces        zone A      $C_{pe} = -1.3$
                   zone B      $C_{pe} = -0.8$

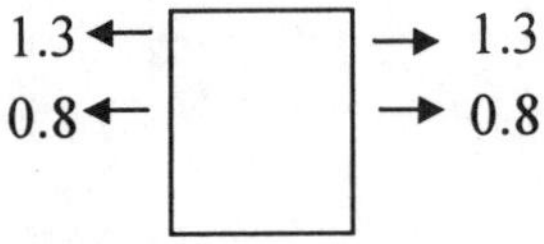

1.3 ←    → 1.3
0.8 ←    → 0.8

**Figure 1.14**

Consider the wind acting on the gable face of the building

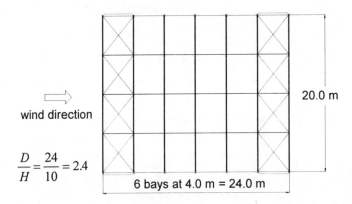

wind direction

$$\frac{D}{H} = \frac{24}{10} = 2.4$$

20.0 m

6 bays at 4.0 m = 24.0 m

**Figure 1.15**

Using *Table 5* and interpolating as before

windward face    $C_{pe} = +0.71$

leeward face     $C_{pe} = -0.21$     $0.71 \rightarrow$ ☐ $\rightarrow 0.21$

**Figure 1.16**

*Clause 2.4.1.3*

$$b \leq \text{crosswind breadth} \quad = \quad 20.0 \text{ m}$$
$$\leq 2 \times \text{height} \qquad = \quad 2 \times 10.0 \quad = 20.0 \text{ m}$$

Use the smaller value of $b$    i.e.    $b = 20.0$ m

*Figure 12(b)* $D = 24.0 > b$ ∴ consider the longitudinal surface as three zones A, B and C.

The width of zone A is equal to $0.2 \times b$   =   $0.2 \times 20$   =   4.0 m
The width of zone B is equal to $0.8 \times b$   =   $0.2 \times 20$   =   16.0 m
The width of zone C is equal to $D - b$   =   $24 - 20$   =   4.0 m

| | | |
|---|---|---|
| zone A | zone B | zone C |

Longitudinal elevation
**Figure 1.17**

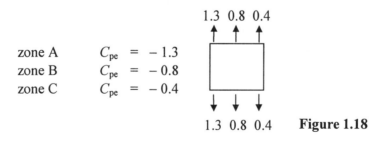

zone A     $C_{pe} = -1.3$
zone B     $C_{pe} = -0.8$
zone C     $C_{pe} = -0.4$

1.3   0.8   0.4

1.3   0.8   0.4     **Figure 1.18**

*Clause 2.5.2.4*     *Table 10*     *Figure 20*
This clause defines the external pressure coefficients for the roof of buildings

*Figure 20*     pitch angle     $\alpha = \tan^{-1}\dfrac{4}{10} = 21.8°$

*Clause 2.5.2.2* this clause defines the loaded zones which are indicated in *Figure 20* and are based on variables $b_L$ and $b_W$
where:

$b_L \leq L$ (crosswind dimension with the wind on the side of the building)
    $\leq 2H$

$b_W \leq W$ (crosswind dimension with the wind on the gable of the building)
$\quad\;\; \leq 2H$
$\quad\;\; L = 24.0 \text{ m} \qquad W = 20.0 \text{ m} \qquad 2H = 20.0 \text{ m} \qquad \therefore\; b_L = b_W = 20.0 \text{ m}$

Consider the wind on the longitudinal face of the building

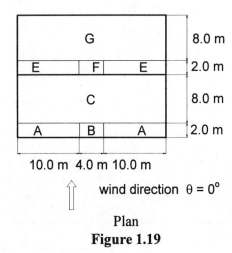

Plan

**Figure 1.19**

In *Table 10* interpolation is required between $+15°$ and $+30°$

| pitch angle | Zone for $\theta = 0°$ | | | | | |
|---|---|---|---|---|---|---|
| | A | B | C | E | F | G |
| 21.8° | − 1.21 | − 0.65 | − 0.25 | − 0.92 | − 0.67 | − 0.45 |
| | + 0.47 | + 0.34 | + 0.29 | | | |

**Extract from Table 10 BS 6399:Part 2 1995**

Both +ve and −ve values are given for zones A, B and C, the most onerous value should be selected when considering combinations with internal pressure coefficients and other load types.

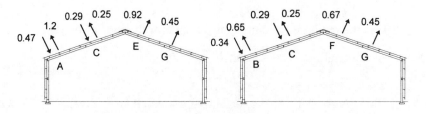

**Figure 1.20**

The zones A, B, E and F are normally used when designing for local effects where high local suction can occur. When calculating the load on entire structural elements such as roofs and walls as a whole, then the values for C and G should be adopted as shown in Figure 1.21

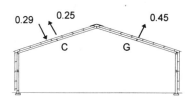

**Figure 1.21**

Consider the wind blowing on the gable of the building

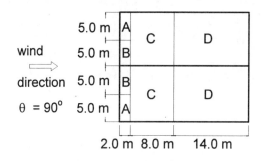

**Figure 1.22**

As before, using interpolation in *Table 10*

| pitch angle | Zone for θ = 0° | | | |
|---|---|---|---|---|
| | A | B | C | D |
| 21.8° | −1.42 | −1.32 | −0.6 | −0.25 |

**Extract from Table 10 BS 6399:Part 2 1995**

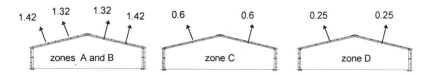

**Figure 1.23**

As before use zones A and B for local effects and zone C which is more onerous than D for entire structural elements.

### Internal Pressure Coefficients: *Clause 2.6  Table 16  Table 17*
The internal pressure coefficients are given in *Clause 2.6* relating to enclosed buildings and buildings with dominant openings. A dominant opening is defined as one in which its area is equal to, or greater than, twice the sum of the openings in other faces which contribute porosity to the internal volume containing the opening. In cases when dominant

openings occur they will control the internal pressure coefficients and should be determined using *Table 17*; in other cases *Table 16, Clauses 2.6.1.1* and *2.6.1.2* should be used.

In many cases for external walls $C_{pi}$ should be taken as either −0.3 or +0.2, whichever gives the larger net pressure coefficient across the walls.
i.e.

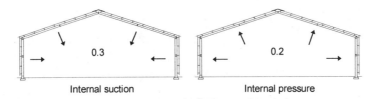

**Figure 1.24**

A summary of the combined external and internal pressure coefficients is shown in Figure 1.25(a) and (b) and Figure 1.26.

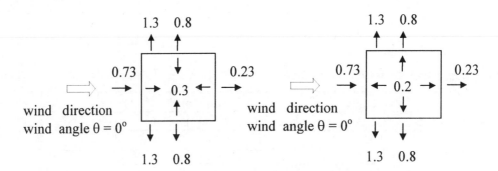

**Figure 1.25(a)**

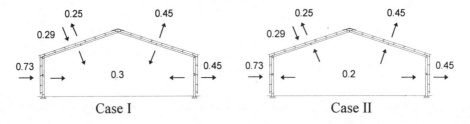

**Figure 1.25(b)**

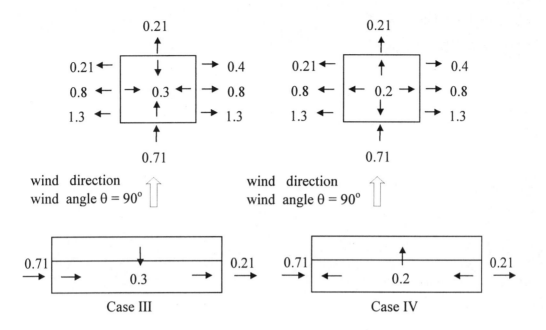

**Figure 1.26**

*Clause 2.1.3.4 Figure 4 Figure 5*
The size effect factor ($C_a$) is dependent on the diagonal dimension '$a$' as defined in *Figure 5* for external pressures and in *Clause 2.6* for internal pressures.

Diagonal dimension for external pressures

| | | | |
|---|---|---|---|
| *Figure 5* | considering the longitudinal walls | $a = \sqrt{6^2 + 24^2}$ | $\approx 25$ |
| | considering the gables | $a = \sqrt{6^2 + 20^2}$ | $\approx 21$ |
| | considering the roof | $a = \sqrt{4^2 + 10^2 + 24^2}$ | $\approx 26$ |

Diagonal dimension for internal pressures

*Clause 2.6.1.1*

$$a = 10 \times \sqrt[3]{\text{int ernal volume of storey}}$$
$$a = 10 \times \sqrt[3]{(20 \times 24 \times 6) + (0.5 \times 20 \times 4 \times 24)} \approx 157$$

Size factors ($C_a$)
*Figure 4*   Site in country   closest distance to sea = 2.0 km   $H_e = 10.0$ m
Use *line A* in graph to determine $C_a$

Considering external pressures

| Size factor for | longitudinal walls | $C_a \approx 0.9$ |
| | gables | $C_a \approx 0.91$ |
| | roof | $C_a \approx 0.88$ |

Considering internal pressures

Size factor for all surfaces $\qquad C_a \approx 0.79$

Consider Case I
*Clause 2.1.3.1 and 2.1.3.3*

windward wall

$$p_e = q_s C_{pe} C_a = 1.07 \times 0.73 \times 0.9 = 0.7 \text{ kN/m}^2$$
$$p_i = q_s C_{pi} C_a = 1.07 \times -0.3 \times 0.79 = -0.25 \text{ kN/m}^2$$

net surface pressure $\quad p = (p_e - p_i) = (0.7 + 0.25) = \mathbf{0.95\ kN/m^2} \longrightarrow$

windward roof slope

$$p_e = q_s C_{pe} C_a = 1.07 \times 0.29 \times 0.88 = 0.27 \text{ kN/m}^2$$
$$p_i = q_s C_{pi} C_a = 1.07 \times -0.3 \times 0.79 = -0.25 \text{ kN/m}^2$$

net surface pressure $\quad p = (p_e - p_i) = (0.27 + 0.25) = \mathbf{0.52\ kN/m^2}\ \searrow$

leeward wall

$$p_e = q_s C_{pe} C_a = 1.07 \times -0.23 \times 0.9 = -0.22 \text{ kN/m}^2$$
$$p_i = q_s C_{pi} C_a = 1.07 \times -0.3 \times 0.79 = -0.25 \text{ kN/m}^2$$

net surface pressure $\quad p = (p_e - p_i) = (-22 + 0.25) = \mathbf{0.03\ kN/m^2}\ \longleftarrow$

leeward roof slope

$$p_e = q_s C_{pe} C_a = 1.07 \times -0.45 \times 0.88 = -0.42 \text{ kN/m}^2$$
$$p_i = q_s C_{pi} C_a = 1.07 \times -0.3 \times 0.79 = -0.25 \text{ kN/m}^2$$

net surface pressure $\quad p = (p_e - p_i) = (-0.42 + 0.25) = \mathbf{0.17\ kN/m^2}\ \nearrow$

Gables  zone A

$$p_e = q_s C_{pe} C_a = 1.07 \times -1.3 \times 0.91 = -1.27 \text{ kN/m}^2$$
$$p_i = q_s C_{pi} C_a = 1.07 \times -0.3 \times 0.79 = -0.25 \text{ kN/m}^2$$

net surface pressure $\quad p = (p_e - p_i) = (-1.27 + 0.25) = \mathbf{1.02\ kN/m^2}\ \downarrow$

Gables  zone B

$$p_e = q_s C_{pe} C_a = 1.07 \times -0.8 \times 0.91 = -0.78 \text{ kN/m}^2$$
$$p_i = q_s C_{pi} C_a = 1.07 \times -0.3 \times 0.79 = -0.25 \text{ kN/m}^2$$

net surface pressure $\quad p = (p_e - p_i) = (-0.78 + 0.25) = \mathbf{0.53\ kN/m^2}\ \downarrow$

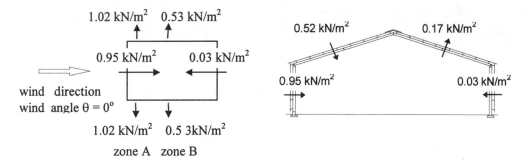

1.02 kN/m²  0.53 kN/m²

0.95 kN/m²    0.03 kN/m²

wind direction
wind angle θ = 0°

1.02 kN/m²   0.5 3kN/m²

zone A   zone B

0.52 kN/m²         0.17 kN/m²

0.95 kN/m²          0.03 kN/m²

**Figure 1.27**

*Clause 2.1.3.5*

The net load on an area of building is given by $P = pA$

A typical internal frame supports surface areas as shown in Figure 1.29

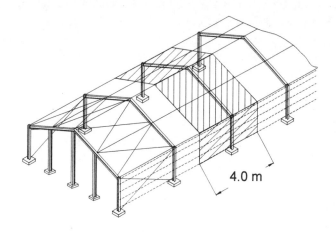

4.0 m

**Figure 1.28**

Area of wall supported/frame   $A = 4.0 \times 6.0 = 24.0 \text{ m}^2$

Area of roof supported/frame   $A = 4.0 \times 10.77 = 43.1 \text{ m}^2$

Surface loads (as shown in Figure 1.29)

   windward wall               $P = 0.95 \times 24.0 = 22.8 \text{ kN}$

   windward roof slope      $P = 0.52 \times 43.1 = 22.41 \text{ kN}$

   leeward wall                 $P = 0.03 \times 24.0 = 0.72 \text{ kN}$

   leeward roof slope        $P = 0.17 \times 43.1 = 7.33 \text{ kN}$

Only Case I is considered here; when designing such a frame all cases must be considered in combination with dead and imposed loads and appropriate partial load factors as given in *Table 2* of BS 5950:Part 2, to determine the critical design load case.

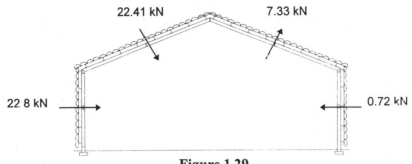

22.41 kN                                      7.33 kN

22.8 kN →                                      ← 0.72 kN

**Figure 1.29**

## 1.12  Example 1.8  Radar reflector

Radar equipment sited on the perimeter of an airfield comprises a revolving reflector
mounted on a tripod trestle as shown in Figure 1.30. Using the data provided determine
the force exerted by the wind on the reflector.

**Data:**
    Location                                              near Aberdeen
    Closest distance to sea                               5 km
    Altitude above mean sea level                         15 m

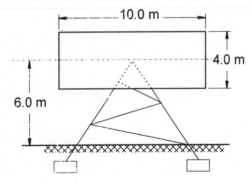
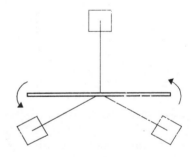

**Figure 1.30**

**Solution:**
*Clause 2.1.3.3 (b)*
When considering canopies, grandstands, open-sided buildings, building elements, free
standing walls, parapets and signboards, the net surface pressure is given by:

$$p = q_s C_p C_a$$

where $C_p$ is the net pressure coefficient for a canopy surface or element and is defined in
*Clauses 2.5.9, 2.7* and *2.8.* In this problem the radar equipment can be treated as indicated
in *Clause 2.8.2;* the wind force on the members of the trestle will be neglected.

$$\therefore C_p = 1.8$$

*Clause 2.1.3.4  Figure 4*
Size effect factor $(C_a)$     $H_e = 8.0$ m,     Site in country, Distance to sea = 5 km

Use *line A* in graph assuming the diagonal dimension '*a*' to be 5.0 m
$$C_a \approx 0.99$$

*Clause 2.2.1    Figure 5*
Location    Aberdeen    Basic wind speed    $V_b = 24$ m/sec

*Clause 2.2.2.2.2*

Altitude factor    $S_a = 1 + (0.001 \times 15)$
$= 1.015$

*Clause 2.2.2*

Site wind speed    $V_s = V_b \times S_a \times S_d \times S_s \times S_p$
Assume    $S_d = S_s = S_p = 1.0$
$V_s = 24.0 \times 1.015 = 24.36$ m/sec

*Clause 2.2.3.3   Table 4*
The values for $S_b$ in *Table 4* can be found by interpolation.
In this example $H_e = 8.0$ m and closest distance to sea is 5 km $\therefore S_b = 1.68$

*Clause 2.2.3.1*
Effective wind speed    $V_e = 24.36 \times 1.68$    $= 40.9$ m/sec

*Clause 2.1.2.1*
Dynamic wind pressure    $q_s = (0.613 \times 40.9^2)/10^3 = 1.03$ kN/m$^2$
Net surface pressure    $p = 1.03 \times 1.8 \times 0.99 = 1.84$ kN/m$^2$

*Clause 2.1.3.5*
Net load on the reflector    $P = 1.84 \times (10.0 \times 4.0) = 73.6$ kN

This force should be considered to be acting at the mid-height of the reflector and within the middle half as shown in Figure 1.31.

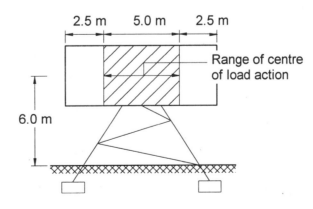

**Figure 1.31**

# 2. Flexural Members

## 2.1 Introduction

The most frequently used, and possibly the earliest used, structural element is the beam. The primary function of a beam is to transfer vertical loading to adjacent structural elements such that the load can continue its path through the structure to the foundations. Loading can be imposed on a beam from one or several of a number of sources, e.g. other secondary beams, columns, walls, floor systems or directly from installed plant and/or equipment. In most cases static loading will be considered the most appropriate for design purposes, but dynamic and fatigue loading may be more critical in certain circumstances.

The structural action of a beam is predominantly bending, with other effects such as shear, bearing and buckling also being present. In addition to ensuring that beams have sufficient strength capacities to resist these effects, it is important that the stiffness properties are adequate to avoid excessive deflection or local buckling of the cross-section (see Section Classification 2.2). A large variety of cross-sections are available when selecting a beam for use in any one of a wide range of applications. The most common types of beam with an indication of the span range for which they may be appropriate are given in Table 2.1. For lightly loaded and small spans such as roof purlins and side sheeting rails the use of hot-rolled angle sections or channel sections is appropriate (see Figure 4.11). Cold-formed sections pressed from thin sheet and galvanised and provided by proprietary suppliers are frequently used. In small to medium spans hot-rolled joists, universal beams (UBs), hollow sections and UBs with additional welded flange plates (compound beams) are often used. If the span and/or magnitude of loading dictates that larger and deeper sections are required, castellated beams formed by welding together profiled cut UB sections, plate girders or box girders in which the webs and flanges are individual plates welded together can be fabricated. Plate girder design is discussed in more detail in Chapter 6. While careful detailing can minimize torsional effects, when they are considered significant hollow tube sections are more efficient than open sections such as UBs, universal columns (UCs), angles and channels.

The section properties of all hot-rolled sections and cold-formed sections are published by their manufacturers; those for fabricated sections must be calculated by the designer (Ref.:11)

The span of a beam is defined in *Clause 4.2.1.1* of BS 5950:Part 1 as the distance between points of effective support. In general, unless the supports are wide columns or piers then the span can be considered as the centre-to-centre of the actual supports or columns.

The most widely adopted section to be found in building frames is the Universal Beam. The design of beams to satisfy the requirements of BS 5950:Part 1 includes the consideration of:

- ♦ section classification,
- ♦ shear capacity,

- ◆ moment capacity (including lateral torsional buckling),
- ◆ deflection,
- ◆ web buckling
- ◆ web bearing,
- ◆ torsional capacity (not required for the design of most beams)

All of these criteria are explained and considered separately and illustrated in Examples 2.1 to 2.10. Examples 2.11 and 2.12 illustrate the design of beams using all criteria (except torsional capacity).

**Table 2.1**

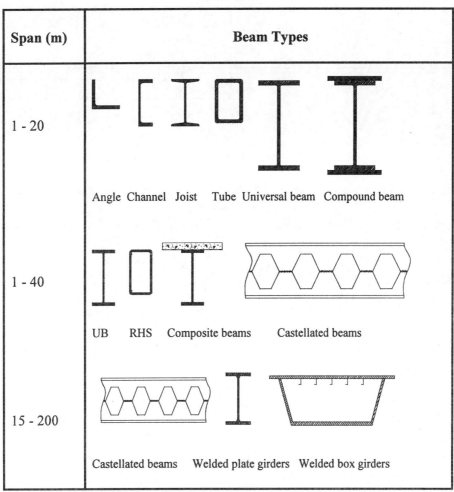

| Span (m) | Beam Types |
|---|---|
| 1 - 20 | Angle  Channel  Joist  Tube  Universal beam  Compound beam |
| 1 - 40 | UB  RHS  Composite beams  Castellated beams |
| 15 - 200 | Castellated beams  Welded plate girders  Welded box girders |

Section Classification *(Clause 3.5)*
In *Clause 3.5* of BS 5950, the cross-sections of structural members are classified into four categories:

- ◆ Class 1      **Plastic** Sections
- ◆ Class 2      **Compact** Sections
- ◆ Class 3      **Semi-compact** Sections
- ◆ Class 4      **Slender** Sections

This classification is based on two criteria:

(i)     the aspect ratio of the elements of a cross-section such as the UB indicated in Figure 2.1, which influences their behaviour when subject to either pure compression, compression caused by bending or a combination of both.

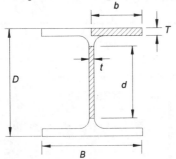

| Element | Aspect ratio |
|---------|--------------|
| outstand of compression flange | b/T |
| web | d/t |

**Figure 2.1**

Reference should be made to *Figure 3* in BS 5950:Part 1 for other cross-sections

(ii)    the moment-rotation characteristics of such a cross-section are as indicated in Figure 2.2.

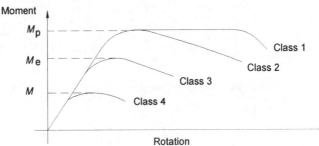

**Figure 2.2**

where:
$M_p$ = plastic moment of resistance
$M_e$ = limiting elastic moment of resistance
$M$ = elastic moment of resistance

These criteria determine whether or not a fully plastic moment as shown in Figure 2.3 can develop within a beam and possess sufficient rotational capacity to permit redistribution of the moments in a structure.

In a beam subject to an increasing bending moment, the bending stress diagram changes from a linearly elastic condition with extreme fibre stresses less than the design strength, ($p_y$), to one in which all of the fibres can be considered to have reached the design strength as shown in Figure 2.3.

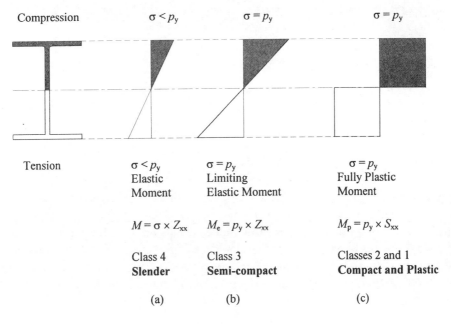

Compression     $\sigma < p_y$     $\sigma = p_y$     $\sigma = p_y$

Tension     $\sigma < p_y$   $\sigma = p_y$     $\sigma = p_y$

Elastic    Limiting     Fully Plastic

Moment    Elastic Moment    Moment

$M = \sigma \times Z_{xx}$    $M_e = p_y \times Z_{xx}$    $M_p = p_y \times S_{xx}$

Class 4     Class 3     Classes 2 and 1

**Slender**     **Semi-compact**     **Compact and Plastic**

(a)     (b)     (c)

**Figure 2.3**

where:

$Z_{xx}$ = elastic section modulus
$S_{xx}$ = plastic section modulus
$\sigma$ = elastic stress
$p_y$ = design strength

**Note:** The **Shape Factor** of a section is defined as:

$$v = \frac{\text{plastic modulus}}{\text{elastic modulus}} = \frac{S_{xx}}{Z_{xx}}$$ The value of $v$ for most I-sections $\approx 1.15$

### 2.1.1 Plastic Sections

The failure of a structure such that plastic collapse occurs is dependent on a sufficient number of plastic hinges developing within the cross-sections of the members (i.e. value of internal bending moment reaching $M_p$), to produce a mechanism. For full collapse this requires one more than the number of redundancies in the structure, as illustrated in Figure 2.4.

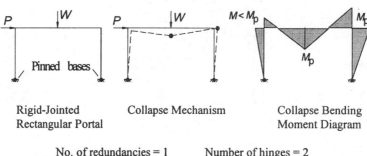

|  Rigid-Jointed | Collapse Mechanism | Collapse Bending |
|  Rectangular Portal | | Moment Diagram |

No. of redundancies = 1          Number of hinges = 2

**Figure 2.4**

The required number of hinges will only develop if there is sufficient rotational capacity in the cross-section to permit the necessary redistribution of the moments within the structure. When this occurs, the stress diagram at the location of the hinge is as shown in Figure 2.3(c), and the aspect ratio of the elements of the cross-section are low enough to prevent local buckling from occurring. Such cross-sections are defined as **plastic sections** and classified as **Class 1.** *Full plastic analysis and design can only be carried out using Class 1 sections.*

### 2.1.2  Compact Sections

When cross-sections can still develop the full plastic moment as in Figure 2.3(c), but are prevented by the possibility of local buckling from undergoing enough rotation to permit re-distribution of the moments, the section is considered to be **compact** and is classified as *Class 2. Compact sections can be used without restricting their capacity, except at plastic hinge positions.*

### 2.1.3  Semi-compact Sections

Semi-compact sections may be prevented from reaching their full plastic moment by local buckling of one or more of the elements of the cross-section. The aspect ratios may be such that only the extreme fibre stress can attain the design strength before local buckling occurs. *Such sections are classified as Class 3 and their capacity is therefore based on the limiting elastic moment as indicated in Figure 2.3(b).*

### 2.1.4  Slender Sections

When the aspect ratio is relatively high, then local buckling may prevent any part of the cross-section from reaching the design strength. *Such sections are called slender sections and are classified as Class 4 sections; their capacity is based on a reduced design strength as specified in Clause 3.6 of BS 5950:Part 1.*

The limiting aspect ratios for elements of the most commonly used cross-sections subject to pure bending, pure axial load or combined bending and axial loads are given in *Table 7* of BS 5950:Part 1. The values given in *Table 7* must be modified to allow for the design strength $p_y$. This is done by multiplying each limiting ratio by $\varepsilon$ which equals

$\left(\dfrac{275}{p_y}\right)^2$. The use of *Clause 3.5* and *Table 7* is illustrated in the examples in this chapter.

## 2.2 Shear Capacity *(Clause 4.2.3)*

Generally in the design of beams for buildings, the effects of shear are negligible and will not significantly reduce the value of the moment capacity. It is evident from the elastic shear stress distribution in an I-beam, as shown in Figure 2.5 that the web of a cross-section is the primary element which carries the shear force.

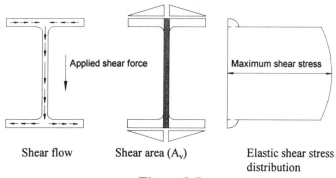

Shear flow      Shear area ($A_v$)      Elastic shear stress distribution

**Figure 2.5**

In situations such as at internal supports of continuous beams where there is likely to be high coincident shear and moment effects which may induce significant principal stresses, (see Figure 2.6), it is necessary to consider the reduction in moment capacity caused by the effects of the shear.

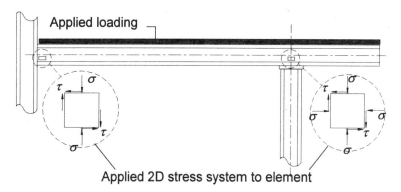

Applied 2D stress system to element

**Figure 2.6**

The shear capacity of a beam is defined in the code as:

$$P_v = 0.6\, p_y\, A_v$$

where:

$0.6\, p_y$ is approximately equal to the yield stress of steel in shear

$A_v$ is the shear area as defined in *Clause 4.2.3*

When the applied ultimate shear force ($F_v$) is equal to or greater than 60% of $P_v$ (i.e. $F_v \geq 0.36\, p_y\, A_v$) the moment capacity of a beam should be reduced as specified in *Clause 4.2.6*.

When the aspect ratio of a web is greater than $63\varepsilon$ the possibility of shear buckling should be considered. This is illustrated in the design of plate girders in Chapter 6. In the design of webs of variable thickness and/or which contain large holes (e.g. castellated beams) the code requires that shear stresses be calculated from first principles assuming elastic behaviour and a maximum shear stress not exceeding $0.7p_y$; the design of such webs is not considered in this text.

## 2.3 Example 2.1 Shear check of a simply supported beam

A simply supported $406 \times 178 \times 74$ UB is required to span 4.5 m and carry an ultimate design load of 40 kN/m. Check the suitability of the section with respect to shear (section properties are given in Ref:11).

**Solution:**

Section properties:  $t = 9.5$ mm,   $D = 412.8$ mm,   $d = 360.4$ mm

Design shear force at the end of the beam $F_v = \dfrac{40 \times 4.5}{2} = 90$ kN

*Clause 4.2.3*                    $P_v = 0.6\, p_y\, A_v$

For a rolled UB section $A_v = tD$

Shear area    $A_v = (9.5 \times 412.8) = 3.922 \times 10^3$ mm$^2$

*Clause 3.1.1*        Web thickness    $t = 9.5$ mm

*Table 6* gives    $p_y = 275$ N/mm$^2$

*Clause 3.5*    Since the beam is subject to pure bending the neutral axis will be at mid-depth.

Section Classification:

*Table 7*        **Note:** $\varepsilon = 1.0 \dfrac{d}{t} = 37.9 < 79\varepsilon$

**Web is plastic**

Shear capacity    $P_v = \dfrac{0.6 \times 275 \times 3.922 \times 10^3}{10^3} = 647$ kN

$>> F_v$ **(90 kN)**

This value indicates the excessive reserve of shear strength in the web.

## 2.4 Moment Capacity *(Clause 4.2.5)*

The moment capacity of a beam is determined by a number of factors such as:

(i)    design strength                     *(Table 6)*
(ii)   section classification              *(Table 7)*
(iii)  elastic section modulus             (Z)
(iv)   plastic section modulus             (S)
(v)    co-existent shear                   *Clauses (4.2.5 and 4.2.6)*    and
(vi)   lateral restraint to the compression flange    *(Clause 4.3)*

The criteria (i) to (v) are relatively straightforward to evaluate, however criterion (vi) is related to the lateral torsional buckling of beams and is much more complex. The design of beams in this text are considered in two categories:

(a)   beams in which the compression flange is fully restrained and lateral torsional buckling cannot occur,    and
(b)   beams in which either no lateral restraint or only intermittent lateral restraint is provided to the compression flange.

### 2.4.1  Compression Flange Restraint

As indicated in Figure 2.3 a beam subject to bending is partly in tension and partly in compression. The tendency of an unrestrained compression flange in these circumstances is to deform sideways and twist about the longitudinal axis as shown in Figure 2.7.

**Figure 2.7**

This type of failure is called *lateral torsional buckling* and will normally occur at a value of applied moment less than the moment capacity ($M_c$) of the section as given in *Clauses 4.2.5* and *4.2.6* and is known as the *buckling resistance moment $M_b$*, defined in *Clause 4.3.7.3.* as:

$$M_b = S_{xx} p_b$$

where:
$S_{xx}$    is the plastic modulus
$p_b$    is the bending strength

The tendency for the compression flange to deform is influenced by:

(i)    lateral restraint,
(ii)   torsional restraint,
(iii)  flange thickness    and
(iv)   effective buckling length.

### 2.4.1.1  *Lateral Restraint*    (*Clause 4.3.2*)

*Full lateral restraint*
It is always desirable where possible to provide full lateral restraint to the compression flange of a beam. The existence of either a cast-in-situ or precast concrete slab which is supported directly on the top flange or cast around it is normally considered to provide adequate restraint. A steel plate floor tack-welded or bolted to the flange also provides adequate restraint; steel floors which are fixed in a manner such that removal for access is required are not normally considered adequate for restraint. Timber floors and beams are frequently supported by steel beams. Generally unless they are fixed to the beam by cleats, bolts or some other similar method and are securely held at their remote end or along their lenghth they are **not** considered to provide adequate restraint.

Full lateral restraint is defined in *Clause 4.2.2.* as being present *"......if the frictional or positive connection of a floor or other construction to the compression flange of the member is capable of resisting a lateral force of not less than 2.5% of the maximum factored force in the compression flange of the member under factored loading. This load should be considered as distributed uniformly along the flange......"*, this is illustrated in Figure 2.8.

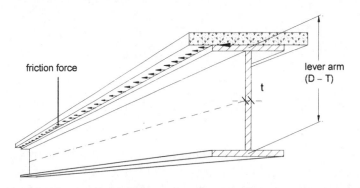

Lateral friction force between underside of slab and top flange

**Figure 2.8**

Minimum friction force required $\qquad = \qquad \dfrac{\text{2.5\% of Applied Moment}}{\text{Lever Arm} \times \text{span}}$

Friction force required $\quad = \quad \dfrac{0.025 \times M}{L(D - T)} \; \text{kN} / \text{m}$

Friction force provided $\quad = \quad \mu \times R$

where:
$M$    maximum moment due to applied factored loads
$L$    span of the beam
$\mu$    coefficient of friction between the concrete beam and the steel flange
$R$    maximum factored vertical load/m applied to the beam
$D$ and $T$ are as before

As discussed above, there is no need to carry out this calculation in the case of concrete floor slabs

*Intermittent lateral restraint*
Most beams in buildings which do not have full lateral restraint are provided with intermittent restraint in the form of secondary beams, ties or bracing members as shown in Figure 2.9

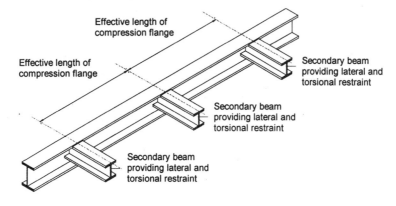

**Figure 2.9**

It is important to ensure that the elements providing restraint are an integral part of a braced structural system and are capable of transmitting the lateral force of 2.5% described previously, divided between the intermediate lateral restraints in proportion to their spacing, (see *Clause 4.3.2.1*). If three or more intermediate lateral restraints are present each individual restraint must be capable of resisting at least 1% of the flange force.

2.4.1.2  *Torsional restraint*      (*Clause 4.3.3*)

A beam is assumed to have torsional restraint about it's longitudinal axis at any location where both flanges are held in their relative positions by external members during bending; as illustrated in Figure 2.10.

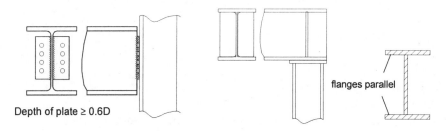

Beam with torsional restraint

**Figure 2.10**

This type of restraint may be provided by load bearing stiffeners as described in *Clause 4.5.8* or by the provision of adequate end connection details as discussed in Chapter 5.

*Beam without torsional restraint*
In situations where a beam is supported by a wall as in Figure 2.11, no torsional restraint is provided to the flanges and buckling is more likely to occur.

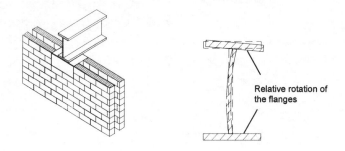

**Figure 2.11**

**2.4.2  *Effective Length L_e***      (*Clause 4.3.5 and 4.3.6*)

The provision of lateral and torsional restraints to a beam introduces the concept of **effective length**. The effective length of a compression flange is the equivalent length between restraints over which a pin-ended beam would fail by lateral torsional buckling. The values to be used in assessing this are given in *Tables 9* and *10* for beams and cantilevers respectively. The values adopted depend on three factors relating to the degree of lateral and torsional restraint at the position of the intermittent restraints, they are:

(a)    the existence of torsional restraints

(b)    the degree of lateral restraint of the compression flange

(c)    the type of loading.

In the case of beams factors (a) and (b) give rise to five possible conditions.

**(a)**    When full torsional restraint exists:

    (i)    both the compression and tension flanges are fully restrained against rotation on plan,

    (ii)   both flanges are partially restrained against rotation on plan, or

    (iii)  both flanges are free to rotate on plan.

**(b)**    When both flanges are free to rotate on plan and the compression flange is unrestrained:

    (i)    torsional restraint is provided soley by connection of the tension flange to the supports,

    (ii)   torsional restraint is provided soley by dead bearing of the tension flange on supports.

Similar conditions exist in *Table 10* for cantilevers.

**(c)**    Type of loading:

A beam load is considered **normal** unless both the beam and the load are free to deflect laterally and so induce lateral torsional buckling by virtue of the combined freedom; in which case the load is a **destabilising** load. In an efficiently designed braced structural system destabalising loads should not normally arise. In some instances the existence of such a load is unavoidable e.g. the sidesway induced in crane-gantry girders by the horizontal surge loads, (see Figure 2.12).

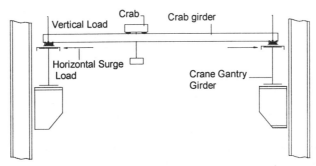

**Figure 2.12**

In *Table 10* it can be seen that destabilising loads are a particular problem for cantilevers since it may be difficult to achieve torsional rigidity at either the free or fixed end. In addition the bottom flange, which is in compression, may not be as readily restrained as the top flange.

### 2.4.3  *Moment Capacity ($M_c$) of Beams with Full Lateral Restraint*

The moment capacity ($M_c$) of beams with the compression flange fully restrained is determined using the following equations which are given in *Clauses 4.2.5* and *4.2.6* for low coincident shear and high coincident shear respectfully.

*Plastic and Compact Sections*
*Clause 4.2.5*     **Low Shear**          $(F_v \le 0.6P_v)$
$$M_c = p_y S$$
$$\le 1.2 p_y Z$$

When the shape factor of a section is greater than 1.2 then $p_y S > 1.2 p_y Z$ and the elastic limitation will govern the design. Generally this will only affect hollow sections and a few Universal Column section. This limitation is to ensure that plasticity does not occur at working loads. There is provision in this clause for enhancing the elastic limitation by replacing the 1.2 factor by the average load factor which equals:

$$\frac{\text{factored loads}}{\text{unfactored loads}} , \text{ this is approximately 1.5}$$

The design of hollow sections, (which tend to have a high shear capacity and high torsional stiffness and do not generally fail by lateral torsional buckling), will be inefficient and uneconomic if this enhancement is not allowed for.

*Clause 4.2.6*     **High Shear**     $(F_v \ge 0.6P_v)$
$$M_c = p_y(S - S_v \rho_1)$$
$$\le 1.2 p_y Z$$

where:

$$\rho_1 = \frac{2.5F_v}{P_v} - 1.5$$

$S_v$ is either
   (i)    the plastic modulus of the shear area for sections with equal flanges        or
   (ii)   plasic modulus of the gross cross-sections minus the plastic modulus of that part of the section remaining after deduction of the shear area

The shear areas in (i) and (ii) are illustrated in Figure 2.13.

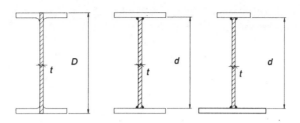

**Figure 2.13**

*Semi - Compact Sections*
*Clauses 4.2.5* and *4.2.6*   **Low Shear and High Shear**
The same value is used for both situations i.e.:

$$M_c = p_y Z$$

*Slender Sections*
*Clauses 4.2.5* and *4.2.6*   **Low Shear and High Shear**
The same value is used for both situations i.e.:

$$M_c = p_y Z$$

where $p_y$ is multiplied by a factor from *Table 8* to allow for the slenderness of the cross-section

The design of beams with fully restrained flanges is normally determined from bending criterion with subsequent checks carried out to ensure that other criteria are satisfied. A beam can be selected such that the applied ultimate moment does not exceed the moment capacity of the section, e.g. for a plastic or compact section assuming the shape factor is less than 1.2:

$$S_{xx} \geq \frac{M_{applied}}{p_y}$$

An appropriate section can be selected from published section property tables.

## 2.5  Example 2.2  Bending in Fully Restrained Beam

A single span beam is simply supported between two columns and carries a reinforced concrete slab in addition to the column and loading shown in Figure 2.14. Using the working loads indicated, select a suitable section considering section classification, shear and bending only. Assume dead loads are inclusive of self-weights.

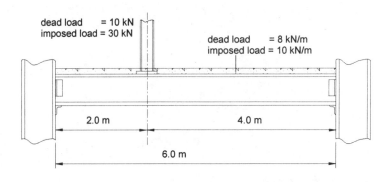

**Figure 2.14**

Solution to Example 2.2 (see Section 2.21)

## 2.6  Moment Capacity *(M_b)* of Beams without Full Lateral Restraint   *(Clause 4.3.7)*

As discussed in Section 2.5.1 the moment capacity of a beam in which the compression flange is not fully restrained is known as the bucklling resistance moment ($M_b$) and is defined in *Clause 4.3.7.3* as

$$M_b = S_{xx}\, p_b$$

The bending strength $p_b$ is dependent on the design strength $p_y$ the equivalent slenderness $\lambda_{LT}$ (equal to *nuvλ* which is defined in Example 2.3) of the section and the type of member i.e. rolled or fabricated by welding. Values of $p_b$ are given in *Tables 11,12* and *19* of the code. Those in *Tables 11* and *12* are for rolled and welded sections respectively and those in *Table 19* are used when using a simplified design procedure which is known as the Conservative Method. A more rigorous and economic design process is illustrated first in this Section. The **Conservative Method** and the use of **Safe Load Tables** are illustrated at the end of the chapter.

### 2.6.1  Rigorous Method      *(Clause 4.3.7.1)*

The condition to be satisfied when considering lateral torsional buckling is:

$$\overline{M} \le M_b$$

where:

$\overline{M}$   is the equivalent uniform moment on any portion of a member between adjacent lateral restraints and is defined in *Clause 4.3.7.2* as *mM_A* in which :

$m$      is an equivalent uniform moment factor determined from *Clause 4.3.7.6*

$M_A$     is the maximum moment on the portion of a member being considered.

## 2.7 Example 2.3 Beam with Intermittent Lateral Restraint

A single span beam of 8.0 m span supporting two factored point loads is shown in Figure 2.15 Assuming lateral and torsional restraint to the compression flange at the ends and points of application of the loads only, check the suitability of a 406 × 140 × 39 UB with respects bending. Neglect self-weight, i.e. the beam is **not** loaded between adjacent lateral restraints.

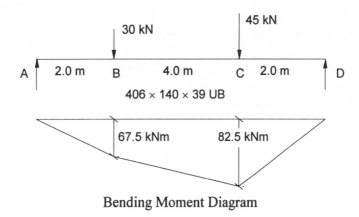

Bending Moment Diagram

**Figure 2.15**

On portion AB     $\overline{M} = m \times 67.5$ kNm
On portion BC     $\overline{M} = m \times 82.5$ kNm
On portion CD     $\overline{M} = m \times 82.5$ kNm

The value of '$m$' may be different for each section.

The evaluation of '$M_A$' is dependent on the structural analysis for the applied load system and resulting bending moment diagram.

The evaluation of '$m$' is determined by reference to *Clause 4.3.7.6, Table 13* and *Table 18* if necessary.

*Clause 4.3.7.6 :* This clause indicates two situations for which to determine '$m$'

    (i) members of uniform cross-section in which case *Table 13* is used to determine '$m$' and '$n$'.

(ii)     members of non-uniform cross-section in which reference should be made to *Appendix B*: beams in this category are not considered in this text.

**Table 13 :**     '*m*' is given the value of 1.0 for any member subject to destabalising loads, any member which is loaded between adjacent restraints and any section with unequal flanges.

In the case of members with equal flanges **not** loaded between adjacent lateral restraints and not subjected to destabalising loads the value of '*m*' is determined from *Table 18*.

The values of '*m*' and '*n*' as determined by *Table 13* requirements are to allow for the effects of a moment gradient along the length of the member. It should be noted from *Table 13* that the two adjustments, '*m*' to the applied bending moment and '*n*' to the slenderness of the compression flange do not occur simultaneously.

**Table 18 :**     the magnitude and sense of the bending moment diagram on the portion of beam being considered defines the value of a factor '*β*' which is susequently used to determine '*m*' where:

$$\beta = \frac{\text{smaller end moment}}{\text{larger end moment}}$$

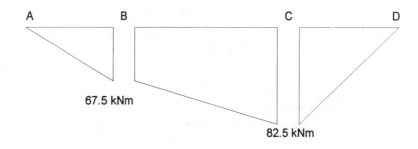

**Figure 2.16**

Portion A-B               Portion B-C               Portion C-D

$$\beta = \frac{0}{67.5} = 0 \qquad \beta = \frac{67.5}{82.5} = 0.82 \qquad \beta = \frac{0}{82.5} = 0$$

*Table 18* values are as follows:

$$m = 0.57 \qquad\qquad m = 0.91 \qquad\qquad m = 0.57$$
$$\overline{M} = 0.57 \times 67.5 \quad \overline{M} = 0.91 \times 82.5 \quad \overline{M} = 0.57 \times 82.5$$
$$= 38.5 \text{ kNm} \qquad = 75.1 \text{ kNm} \qquad = 47.0 \text{ kNm}$$

*Table 18* gives values of $\beta$ in increments of 0.1, the value of '$m$' for intermediate values can be found from the following equation but should not be taken as less than 0.43

$$m = 0.57 + 0.33\beta + 0.1\beta^2 \leq 1.0$$

Clearly, in the above case portion B-C of the beam is critical and a section should be designed such that:

$$M_b \geq 75.1 \text{ kNm}$$

**Equivalent Slenderness Ratio:**     (*Clause 4.3.7.5*)

$$\lambda_{LT} = nuv\lambda$$

where:

$n$   is a slenderness correction factor and is equal to either 1.0, or for members with equal flanges, loaded between adjacent restraints and not subjected to destabilising loads, can be determined from *Tables 15* and *Table 16* (see *Table 13* as for $m$) or for standard load conditions, from *Table 20*.

$u$   is a buckling parameter which for rolled UB, UC or channel section can be taken as 0.9, is more accurately evaluated using *Appendix B*, or is given in published section property tables.

$v$   is a slenderness factor which for uniform, flanged members with at least one axis of symmetry can be determined from *Table 14*.

*Evaluation of '$n$'*

In most cases when using the conservative method in Clause 4.3.7.7, the value of '$n$' can be estimated by modifying the values given in *Table 20*. *Tables 15* and *16* are used for more complex cases, as illustrated in Example 2.4. The value depends on the bending moment diagram of the portion of a beam being considered. Consider the portions of beam ABCD as before and using *Table 20*.

Portion A-B

67.5 kNm      $n = 0.77$

67.5 kNm

82.5 kNm    $n = 1.0$        $n = 0.77$

Portion B-C

$$\text{Use average value} \therefore n = \frac{(1.0 + 0.77)}{2} = 0.89$$

Portion C-D

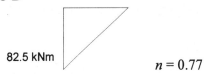

82.5 kNm                                 $n = 0.77$

The critical value of n relates to portion C-B
In Examlpe 2.3, (where no load occurs between the lateral restraints), using the Rigorous
Method the value given in Table 13 for the slenderness correction factor is      **$n = 1.0$**

***Evaluation of 'u'***
From section property tables for a 406 × 140 × 39  UB                            **$u = 0.858$**

***Evaluation of 'v'***
The values of 'v' given in *Table 14* are dependent on three factors $N$, $\lambda$, and $X$.

$N$      this factor equals 0.5 for members with equal flanges or in the case of other sections

equals $\dfrac{I_{cf}}{I_{cf} + I_{tf}}$

where $I_{cf}$ and $I_{tf}$ are the second moments of area of the compression and tension flanges
respectively about the y-y axis of the section.

$\lambda$  is the minor axis slenderness and is defined as $\lambda = \dfrac{L_e}{r_{yy}}$

where:
$L_e$ is the effective length as discussed in Section 2.4.2 and $r_{yy}$ is the radius of gyration
about the minor axis of the member; normally from the section properties tables.

$X$  is the torsional index and may be determined from *Appendix B*, section property tables
or taken as $D/T$ for rolled sections, provided that $U$ is taken as 0.9 or 1.0 for other
sections.

*Clause 4.3.7.5*
UB section is being considered          $N$  =  0.5
Full restraint to the compression flange is not provided
*Table 9*   effective length          $L_E$  =  1.0 × 4.0 = 4.0 m
radius of gyration                      $r_{yy}$ =  28.7

slenderness                             $\dfrac{L_e}{r_{yy}} = \dfrac{4000}{28.7} = 139.4$

                                                                                **$\lambda = 139.4$**

torsional index                         $X$  =  47.5

                                        $\dfrac{\lambda}{X} = \dfrac{139.4}{47.5} = 2.95$

*Table 14*     $N = 0.5$   and     $\dfrac{\lambda}{X}$ = 2.95           **$v = 0.91$**

equivalent slenderness          $\lambda_{LT}$   =   $nuv\lambda$

                                  $\lambda_{LT}$   =   $1.0 \times 0.858 \times 0.91 \times 139.4 = 108.8$

                                                             **$\lambda_{LT} = 108.8$**

*Table 6*          $T < 16.0$ mm     $p_y$   =   $275$ N/mm$^2$

*Clause 4.3.7.4 Table 11*          $\lambda_{LT}$   =   $108.8$           **$p_b$ = 111N/mm$^2$**

$$M_b = S_{xx}\, p_b = (724 \times 10^3 \times 111) / 10^6 = 80.36 \text{ kNm}$$
$$\overline{M} < M_b$$

**Section is adequate in bending**

## 2.8 Example 2.4 Beam with intermittent lateral restraint

A 406 × 178 × 54 UB is simply supported and carries factored loading as shown in Figure 2.17. Assuming lateral restraints to the compression flange at A, B, C and D as in Example 2.3, check the suitability of the section with respect to bending.

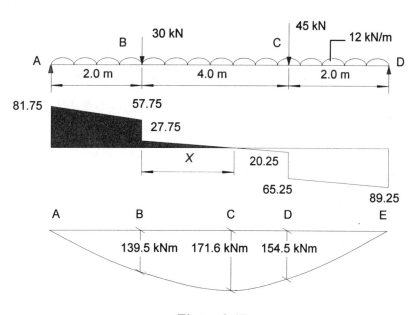

**Figure 2.17**

Distance to point of zero shear from B            $x$ = $\dfrac{27.75}{12}$ = 2.31 m

Position of maximum bending moment is 4.31 m from the left-hand end.

Bending moment at B $= \dfrac{(81.75+57.75)2}{2} \quad = \quad 139.5$ kNm

Bending moment at C $= \dfrac{(65.25+89.25)2}{2} \quad = \quad 154.5$ kNm

Max. bending moment at 2.3 m $= \dfrac{(81.75+57.75)2}{2} + \dfrac{(2.31 \times 27.75)}{2} \quad = \quad 171.6$ kNm

Portion BC is the critical section $\qquad \overline{M} = m \times 171.6$ kNm

*Clause 4.3.7.6*      *Table 13*              $m = 1.0$

$$\overline{M} = \mathbf{171.6 \ kNm}$$

The value of '$n$' is determined from *Tables 15* and *16*

*Table 15* applies to members in which the applied loading is substantially concentrated within the middle fifth of the unrestrained length; *Table 16* applies to all other load cases.

   In this example for portion BC *Table 16* is appropriate and the value of '$n$' is dependent on two variables

$$\text{(i)} \qquad \gamma = M/M_o$$

where:
$M$   is the greater end moment on the portion of beam being considered,
$M_o$   is the mid-span moment on a simply supported span equal to the unrestrained
        length.
The signs and values to be adopted for $M$ and $M_o$ are illustrated in *Table 17*.

$$\text{(ii)} \quad \beta = \frac{\text{smaller end moment}}{\text{larger end moment}} \text{; as before}$$

Consider portion BC

$$M_o = \frac{12 \times 4.0^2}{8} \qquad\qquad = \ 24 \text{ kNm}$$

$$\gamma = +\frac{M}{M_o} = \frac{154.5}{24} \quad = \ 6.43$$

$$\beta = +\frac{139.5}{154.5} \qquad\qquad = \ 0.9$$

*Table 16* Slenderness correction factor '$n$'

$$n = \mathbf{0.99}$$

From section property tables for a 406 × 178 × 54 UB $\qquad\qquad u = \mathbf{0.871}$

| | | | |
|---|---|---|---|
| *Table 9* | effective length | $L_E = 1.0 \times 4.0 = 4.0$ m | |
| | radius of gyration | $r_{yy} = 38.5$ | |

$$\frac{L_E}{r_{yy}} = \frac{4000}{38.5} = 104 \qquad \lambda = 104$$

torsional index $\qquad X = 38.3$

$$\frac{\lambda}{X} = \frac{104}{38.3} = 2.72$$

*Table 14* $\qquad N = 0.5 \qquad$ and $\qquad \dfrac{\lambda}{X} = 2.72 \qquad\qquad$ **v = 0.92**

Equivalent slenderness $\qquad \lambda_{LT} = nuv\lambda$

$$\lambda_{LT} = 0.99 \times 0.871 \times 0.92 \times 104 = 82.5$$
$$\lambda_{LT} = 82.5$$

*Table 6* $\qquad T < 16.0$ mm $\qquad p_y = 275$ N/mm$^2$

*Clause 4.3.7.4 Table 11* $\qquad \lambda_{LT} = 82.5 \qquad\qquad$ **$p_b = 159.5$ N/mm$^2$**

$$M_b = S_{xx}\, p_b = (1055 \times 10^3 \times 159.5) / 10^6 = 168 \text{ kNm}$$
$$M_b < \overline{M}$$

**Section is inadequate**

Clearly, this section has insufficient bending capacity to support the load and either a larger section or a higher grade of steel should be used.
Assume grade 50 steel

*Table 6* $\qquad T < 16.0$ mm $\qquad p_y = 355$ N/mm$^2$

*Clause 4.3.7.4 Table 11* $\qquad \lambda_{LT} = 82.5 \qquad\qquad$ **$p_b = 182.5$ N/mm$^2$**

$$M_b = S_{xx}\, p_b = (1055 \times 10^3 \times 182.5) / 10^6 = 192.5 \text{ kNm}$$
$$M_b > \overline{M}$$

**Section is adequate using grade 50 steel**

**Note:** In *Clause 4.3* of BS 5950:Part 1, it is stated that:
*'All beams should satisfy the requirementsof 4.2.1 and 4.2.3 to 4.2.6 inclusive.'*
This implies that although $p_b$ is always $\leq p_y$ the value of the buckling moment of resistance $M_b$ may be greater than $1.2p_yZ_{xx}$, as given in *Clauses 4.2.5* and *4.2.6*, and should be checked; the smaller value indicates the design strength. This will only affect a few of the very heavy UC sections or situations in which bending occurs about the y-y axis.

### 2.9  Example 2.5  Rectangular hollow section as a beam

A single span beam supports the 'service' loads indicated in Figure 2.18 and is restrained at points ABCD as in Examples 2.3 and 2.4. Check the suitability of a $300 \times 200 \times 8$ RHS section with respect to bending.

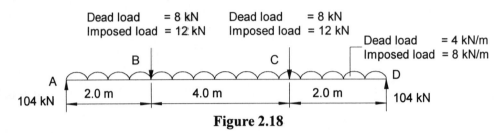

**Figure 2.18**

| | | | |
|---|---|---|---|
| Design load at B and C | = | $(1.4 \times 8) + 1.6 \times 12)$ | = 30.4 kN |
| Design distributed load | = | $(1.4 \times 4) + (1.6 \times 8)$ | = 18.4 kN/m |
| Total factored load | = | $30.4 + 30.4 + (18.4 \times 8)$ | = 208 kN |
| Total unfactored load | = | $20 + 20 + (12 \times 8)$ | = 136 kN |

$$\text{Average load factor} \quad = \quad \frac{208}{136} = 1.53$$

The average load factor will be used to modify the $1.2p_yZ$ value in *Clause 4.2.5*, since this is the governing equation for rolled hollow sections, i.e.

$$M_{cx} \quad = \quad 1.53p_yZ_{xx}$$

Bending moment at B = $(104 \times 2) - (18.4 \times 2)(1)$ = 171.2 kNm
Bending moment at C = 171.2 kNm

$$\text{Bending moment at mid-span} \quad = \quad (30.4 \times 2) + \left( \frac{18.4 \times 8^2}{8} \right) \quad = 208 \text{ kNm}$$

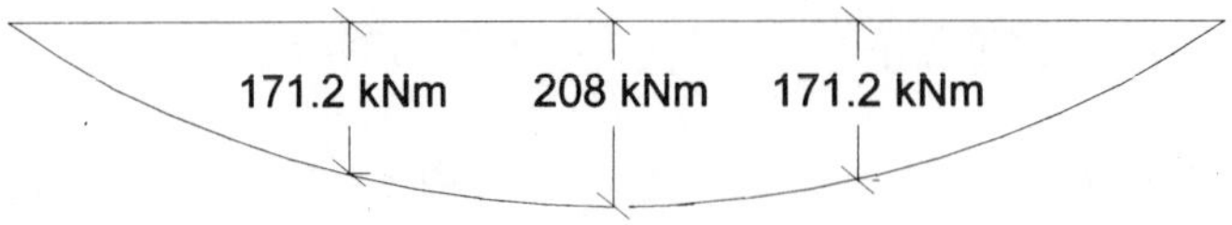

171.2 kNm        208 kNm        171.2 kNm

Bending moment diagram

**Figure 2.19**

*Clause 4.3.7.6  Table 13*      $m = 1.0$              $\overline{M} = 208$ kNm

*Appendix B  Clause B.2.6*      Buckling moment resistance of box sections

This clause states *'Box sections of uniform wall thickness need not be checked for lateral torsional buckling effects provided that* $\lambda = (L_e/r_y)$ *is not greater than the limiting values given in Table 38'*; i.e. when

| *D/B* | $\lambda$ |
|-------|-----------|
| 1 | $\infty$ |
| 2 | $\dfrac{350 \times 275}{p_y}$ |
| 3 | $\dfrac{225 \times 275}{p_y}$ |
| 4 | $\dfrac{170 \times 275}{p_y}$ |

From section property tables

| | | | | |
|---|---|---|---|---|
| *Table 6* | $T < 16$ mm | $p_y$ | = | $275$ N/mm$^2$ |
| *Table 9* | effective length | $L_e$ | = | $1.0 \times 4.0 = 4.0$ m |
| | radius of gyration | $r_{yy}$ | = | $82.3$ mm |

$$\text{slenderness} \quad \lambda = \frac{L_E}{r_{yy}} = \frac{4000}{82.3} = 48.6$$

$$D/B = \frac{300}{200} = 1.5$$

The limiting value of $\lambda$ for $D/B = 2.0$ is given by $\dfrac{350 \times 275}{p_y} = 350$

The critical span above which this section must be checked is given by:

$$350 \times r_{yy} \quad = \quad 350 \times 82.3 \quad = \quad 28805 \text{ mm} \quad \text{i.e. } \mathbf{28.8 \ m}$$

Lateral torsional buckling is clearly not critical and the moment capacity is governed by *Clause 4.2.5*    (**Note**: low shear occurs in this case).

$$\begin{aligned} M_c &= 1.53 p_y Z_{xx} \\ &= (1.53 \times 275 \times 653 \times 10^3)/10^6 \quad = 274.7 \text{ kNm} \end{aligned}$$

$$M_c > 208 \text{ kNm}$$

**Section is adequate**

## 2.10  Example 2.6    Cantilever beam

Consider a beam with a cantilever overhang and supporting the factored loads shown in Figure 2.20.

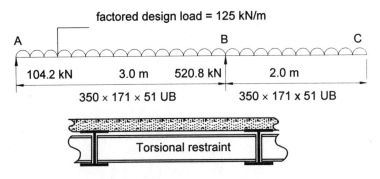

Restraint condition at end of cantilever

**Figure 2.20**

A maximum bending moment and a maximum shear force always occur at the same cross-section in a cantilever. It is essential in these circumstances to ensure that a check is carried out on the effect of the shear force on the moment capacity i.e. either Low Shear with $F_v < 0.6P_v$ in which case *Clause 4.2.5* governs or High Shear with $F_v > 0.6P_v$, in which case *Clause 4.2.6* is used.

Check the suitability of a $356 \times 171 \times 51$ UB with respect to combined shear and bending.

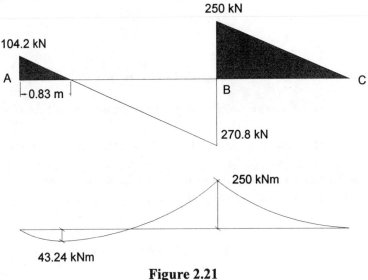

**Figure 2.21**

At position B

| | | |
|---|---|---|
| Design shear force | = | 270.8 kN |
| Design bending moment | = | 240 kNm |

Section properties:

$D = 355$ mm          $B = 171.5$ mm          $T = 11.5$ mm

$r_{yy} = 39.11$ mm     $S_{xx} = 1010 \times 10^3$ mm$^3$     $t = 7.4$ mm

$X = 28.8$          $Z_{xx} = 896 \times 10^3$ mm$^3$     $u = 0.882$

Shear check:     *Clause 4.2.3*

$$P_v = 0.6 p_y A_v = (0.6 \times 275 \times 7.4 \times 355)/10^3 = 433.5 \text{ kN}$$

$$F_v/P_v = \frac{270.8}{433.5} = 0.62 > 0.6 \therefore \text{ High shear}$$

Bending check  *Clause 4.2.6*

Since $F_v > 0.6 P_v$  $\qquad M_c = p_y (S - S_v \rho_1)$
$$\leq 1.2 \, p_y Z$$

where:

$$\rho_1 = \frac{2.5 F_v}{P_v} - 1.5$$

$S_v$    is the plastic modulus of the shear area. For a UB section  $S_v = \dfrac{tD^2}{4}$

$$\rho = \frac{2.5 \times 270.8}{433.5} - 1.5 = 0.06 \qquad\qquad S_v = \frac{7.4 \times 355^2}{4} = 233.15 \times 10^3 \text{ mm}^3$$

$$\begin{array}{llll} p_y (S - S_v \rho_1) &= 275 \times [1010 - (233.15 \times 0.06)]10^3/10^6 &= & 273.9 \text{ kNm} \\ 1.2 p_y Z &= (1.2 \times 275 \times 896 \times 10^3)/10^6 &= & 295.7 \text{ kNm} \\ & \qquad\qquad M_c = 273.9 \text{ kNm} > M_x \end{array}$$

**$M_c$ is adequate**

Lateral torsional buckling check:     *Clause 4.3.7*

| *Clause 4.3.7.1* | | $\overline{M}$ | $\leq$ | $M_b$ |
|---|---|---|---|---|
| *Clause 4.3.7.2* | | $\overline{M}$ | $=$ | $m_A$ |
| *Table 13* | $m = 1.0$ | $\overline{M}$ | $=$ | 250 kNm |
| *Clause 4.3.7.3* | | $M$ | $=$ | $S_{xx} p_b$ |
| *Clause 4.3.7.5* | | $\lambda_{LT}$ | $=$ | $nuv\lambda$ |

*Clause 4.3.6.2* and    *Table 10*

The bottom flange is the compression flange in the cantilever. Assuming the cantilever to be continuous with torsional restraint only at the end, then:

effective length $L_E$    $= 2.4L = 2.4 \times 2000 = 4800 \text{ mm}$

slenderness    $\lambda = \dfrac{4800}{39.1} = 122.8$

*Table 16*    $\beta = 0$

$M = 250 \text{ kNm}$  $\qquad M_0 = \left( 125 \times 1.0 \times \dfrac{1}{2} \right) = 62.5 \text{ kNm}$  $\qquad \gamma = \dfrac{M}{M_0} = \dfrac{250}{62.5} = 4$

**$n = 0.87$**

*Table 14*    $N = 0.5$  $\qquad\qquad \dfrac{\lambda}{X} = \dfrac{122.8}{28.8} = 4.26$

**$v = 0.85$**

$$\lambda_{LT} \quad = \quad 0.87 \times 0.882 \times 0.85 \times 122.8$$

**$\lambda_{LT} = 80.1$**

*Table 11*    $p_y \quad = \quad 275 \text{ N/mm}^2 \qquad \lambda_{LT} \quad = \quad 80.1$

**$p_b = 180.8 \text{ N/mm}^2$**

$$M_b \quad = \quad (1010 \times 10^3 \times 180.8)/10^6 \quad = \quad 182.6 \text{ kNm}$$

$$M_b \ll \overline{M}$$

**Section is inadequate in bending**

This UB section does not satisfy the lateral torsional buckling criterion. A larger section, higher grade of steel, additional restraint to the compression flange or a combination of these modifications can be used to solve this problem.

The very high value of loading in this problem demonstrates the reason for most beams being designed for bending with coincident low shear. The design cases in which checks for high shear are necessary are:

(i)     at the supports of cantilevers,
(ii)    at the location of heavy point loads on a span,
(iii)   where very high distributed loads are applied to short spans.

## 2.11  Web Buckling and Web Bearing

In addition to shear failure of a web as discussed in section 2.3, there are two other modes of failure which may occur, they are:

(i)     web buckling       and
(iii)   web bearing.

At locations of heavy concentrated loads such as support reactions or where columns are supported on a beam flange, additional stress concentrations occur in the web. This introduces the possibility of the web failing in a buckling mode similar to a vertical strut, or by localised bearing failure at the top of the root fillet, as shown in Figure 2.22.

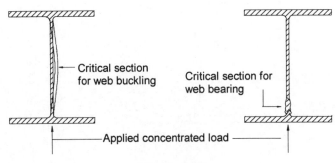

**Figure 2.22**

The code specifies two local capacities relating to these modes of failure. When either of these is less than the applied concentrated force it will be necessary to provide additional strength to the web. In most cases this requires the design of load bearing stiffeners; the detailed design of such stiffeners is given in Chapter 6. There may be other reasons for utilizing stiffeners, such as enhancing torsional stiffness at supports and points of lateral restraint, as discussed previously.

### 2.11.1 Web Buckling    (*Clause 4.5.2*)

In the buckling check the web is considered to be a fixed-end strut between the flanges. It is assumed that the flange through which the concentrated load is applied is restrained against rotation relative to the web and lateral movement relative to the other flange.

The load carrying capacity of a strut is dependent on its compressive strength '$p_c$' and cross-sectional area and is given by:

$$P_w = (b_1 + n_1)tp_c$$

where:

$b_1$      is the stiff bearing length,

$n_1$      is the length obtained by dispersion at 45° through half the depth of the flange,

$t$      is the web thickness     and

$p_c$      is the compressive strength from *Table 27(c)* as shown in Figure 2.23

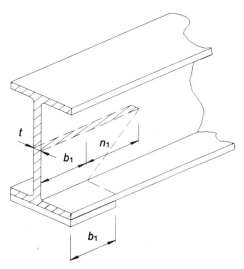

**Figure 2.23**

The slenderness of the strut is given by:        $\lambda = \dfrac{L_E}{r_{yy}}$

$$L_E = 0.7d \quad r_{yy} = \sqrt{\frac{I}{Area}} \quad = \sqrt{\frac{(b_1 + n_1)t^3}{12(b_1 + n_1)t}} \quad = \sqrt{\frac{t^2}{12}} \quad = \frac{t}{2\sqrt{3}}$$

$$\lambda = \frac{0.7d}{\left(\dfrac{t}{2\sqrt{3}}\right)} \quad \approx \quad \frac{2.5d}{t}$$

When a flange is not restrained against rotation and lateral movement, the value of $\lambda$ should be modified by using an effective length based on the values given in *Table 24*.

### 2.11.2  Web Bearing  (Clause 4.5.3)

The bearing check is similar to the buckling check, in that an effective bearing area over which the design strength of the web is assumed to act is determined using:

$$P_{crip} = (b_1 + n_2)t p_{yw}$$

where:
$b_1$      is as before
$n_2$      is the length obtained by dispersion through the flange to the top of the root fillet assuming a slope 1:2.5 to the plane of the flange     and
$p_{yw}$    is the design strength of the web

      as shown in Figure 2.24.

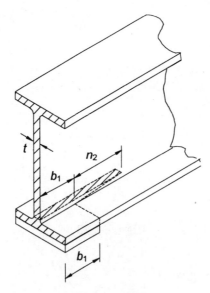

**Figure 2.24**

In both cases at the design stage it is usually necessary to make assumptions regarding the provision of bearing plates at supports or cap plates/base plates on columns to provide stiff bearing. In the code, Clause 4.5.1.3 defines the stiff bearing length $b_1$ as 'that length which cannot deform appreciably in bending'. The value of $b_1$ is determined by assuming a dispersion of load through a bearing plate at 45° and is illustrated in Figure 8 of the code.

## 2.12 Example 2.7 Web bearing and web buckling at support

Consider the beam in Example 2.2 in which the left-hand end reaction is 122.9 kN, and check the suitability of the web with respect to buckling and bearing.

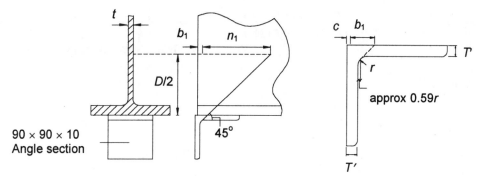

**Figure 2.25 Web buckling**

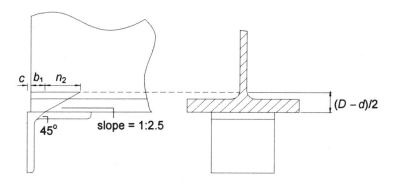

**Figure 2.26 Web bearing**

where:

$c$ is the end clearance,

$b_1 \approx (T' - c) + 0.59r + T' \quad = \quad (2T' + 0.59r - c),$

$n_2 = 2.5 \times \dfrac{(D-d)}{2} \quad\quad = 1.25(D - d),$

$n_1 = D/2$

Assuming $c = 5$ mm

$b_1 = (2 \times 10 + (0.59 \times 11) - 5) = 21.5$ mm

$$n_1 \quad = \quad \frac{306.6}{2} \quad\quad\quad = 153 \text{ mm}$$

$$n_2 \quad = \quad 1.25 \times (306.6 - 265.2) = 52 \text{ mm}$$

Web buckling:

$$P_w = (b_1 + n_1)tp_c$$

Assuming the bottom flange is laterally and torsionally restrained,

$$\lambda \;=\; 2.5d/t \quad = \frac{(2.5 \times 265.2)}{6.7} \;=\; 99$$

*Table 6* $\qquad\qquad\qquad t \;<\; 16 \text{ mm} \quad p_y \;=\; 275 \text{ N/mm}^2 \quad \lambda = 99$

$$\boldsymbol{p_c = 127 \text{ N/mm}^2}$$

$$P_w \;= [(21.5 + 153) \times 6.7 \times 127]/10^3 = \;\; 148.5 \text{ kN}$$
$$P_w \;> 122.9 \text{ kN}$$

**Web buckling strength is adequate**

Web bearing:

$$P_{crip} \;= (b_1 + n_2)tp_{yw}$$
$$P_{crip} \;= [(21.5 + 52) \times 6.7 \times 275]/10^3 \;= 135 \text{ kN}$$
$$P_{crip} \;> 122.9 \text{ kN}$$

**Web bearing strength is adequate**

No web stiffeners are required.

A similar calculation can be carried out at the location of the column on the top flange.

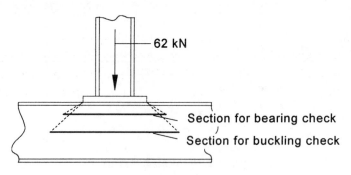

**Figure 2.27**

Since the load is normally less and the distribution in the web is considerably greater, this will generally be less critical than the location at the end reaction. In the case of square and rectangular hollow sections when the flange is not welded to a bearing plate, additional effects of moments induced in the web due to eccentricity of loading as shown in Figure 2.28 must be allowed for.

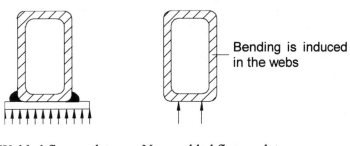

Welded flange plate          Non-welded flange plate

**Figure 2.28**

Reference should be made to the *Structural Design Guide...* (Ref:11, 12) which contains detailed information relating to the relevant bearing factors in such cases.

## 2.13  Deflection of Beams    (*Clause 2.5.1*)

In *Table 1* of the code, one of the serviceability limit states to be considered is deflection. Recommendations for limiting values of deflection under various circumstances are given in *Table 5*.
Limitations on the deflections of beams are necessary to avoid consequences such as:

- damage to finishes, e.g. to brittle plaster, ceiling tiles,
- unnecessary alarm to occupants of a building,
- misalignment of door frames causing difficulty in opening,
- misalignment of crane rails resulting in derailment of crane-gantries.

There are large variations in what are considered by practising engineers to be acceptable deflections for different circumstances. If situations arise in which a designer considers the recommendations given in *Table 5* to be too lenient or too severe (e.g. conflicting with the specification of suppliers or manufacturers) then individual engineering judgement must be used.

The values in *Table 5* relating to beams give a $\dfrac{\text{span}}{\text{coefficient}}$[1] ratio calculated using the service loads only. The coefficient varies from 180 for cantilevers, to 360 for the deflection of beams supporting brittle finishes. In most circumstances, the dead load deflection will have occurred prior to finishes being fixed and the building being in use and will not therefore cause any additional problem while the building is in service. Unfactored loads are used since it is under service conditions that deflection may be a problem.

---

[1] A ratio is used instead of a fixed value since this limits the curvature of the beam which depends on the span.

Additional values are given for portal frames and crane-gantry girders; they are not considered here.

In a simply supported beam, the maximum deflection induced by the applied loading always approximates the mid-span value if it is not equal to it. A number of standard, frequently used load cases for which the elastic deformation is required is given in Table 2.1 of this text. In the case indicated with '*' the actual maximum deflection will be approximately equal to the value given, (i.e. within 2.5%).

In many cases beams support complex load arrangements which do not lend themselves to either an individual or combination of the load cases given in Table 2.1. Since the values in *Table 5* are recommendations for maximum values, approximations in calculating deflection are normally acceptable. Provided that deflection is not the governing design criterion, a calculation which gives an approximate answer is usually adequate. The "Steel Designers Manual" (ref:17), provides a range of coefficients which can be used to either calculate deflections or to determine the minimum I value (second moment of area), to satisfy any particular $\dfrac{\text{span}}{\text{coefficient}}$ ratio.

An *equivalent uniformly distributed load* technique which can be used for estimating actual deflections or required I values for simply supported spans is given in this text.

### *Equivalent UDL technique.*

(a)     Estimating deflection

Consider a single-span, simply supported beam carrying a *non-uniform* loading which induces a maximum bending moment of BMax, as shown in Figure 2.29

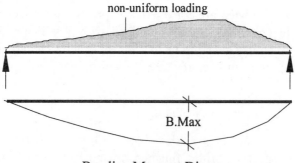

Bending Moment Diagram
**Figure 2.29**

The equivalent U.D.L. ($W_e$) which would induce the same *magnitude* of maximum bending moment (note: the position may be different), on a simply supported span carrying a *uniform* loading can be determined from:

Maximum bending moment = B.Max     $= \dfrac{W_e L^2}{8}$

$$\therefore \qquad W_e = \frac{8\,B.\,\text{Max.}}{L^2}$$

where $W_e$ is the equivalent uniform distributed load

**Table 2.1.**

| Load Case | Maximum Deflection | Load Case | Maximum Deflection |
|---|---|---|---|
| $W_{Total}$ over span $L$ (simply supported) | $\dfrac{5WL^3}{384EI}$ | $W_{Total}$ over span $L$ (fixed ends) | $\dfrac{WL^3}{384EI}$ |
| $W_{Total}$ distributed over $b$, with $a$, $b$, $c$; span $L$ | $\dfrac{WL^3}{384EI}\,\alpha_1$ | $P$ at $L/2$, $L/2$ (fixed ends), span $L$ | $\dfrac{PL^3}{192EI}$ |
| $P$ at $L/2$, $L/2$; span $L$ | $\dfrac{PL^3}{48EI}$ | $P$ at $a$, $b$ ($b > a$), fixed end, span $L$ | $\dfrac{2\,Pa^2b^3}{3\,EI\beta}$ |
| $*$  $P$ at $a$, $b$ ($b > a$); span $L$ | $\approx \dfrac{PL^3}{48EI}\,\alpha_2$ | $W_{Total}$ over $a$, $b$ | $-\dfrac{Wa^2b}{24EI}$ |
| $W_{Total}$ over $a$, $b$; span $L$ | $\dfrac{Wa^3}{8EI}\,\alpha_3$ | $W_{Total}$ over $a$, $b$ | $\dfrac{Wb^4}{8EI}+\dfrac{Wab^3}{6EI}$ |
| $P$ at $a$, $b$ (cantilever), span $L$ | $\dfrac{Pa^3}{3EI}\,\alpha_4$ | $P$ at $a$, $b$ | $\dfrac{Wb^3}{3EI}+\dfrac{Wab^2}{3EI}$ |

$$\alpha_1 = (L^3 + 2L^2a + 4La^2 - 8a^3) \qquad \alpha_2 = \left[\frac{3a}{L} - 4\left(\frac{a}{L}\right)^3\right] \qquad \alpha_3 = \left(1 + \frac{4b}{3a}\right) \qquad \alpha_4 = \left(1 + \frac{3b}{2a}\right)$$

$$\beta = (3L - 2a)^2$$

The maximum deflection of the beam carrying the uniform loading will occur at the mid-span and be equal to
$$\delta = \frac{5W_e L^4}{384 EI}$$
Using this expression, the maximum deflection of the beam carrying the non-uniform loading can be estimated by substituting for the $W_e$ term, i.e.

$$\delta \approx \frac{5W_e L^4}{384 EI} = \frac{5 \times \left(\dfrac{8B.Max}{L^2}\right) L^4}{384\ EI} = \frac{0.104\ B.Max\, L^2}{EI}$$

(b)    Estimating the required second moment of area ($I$) value
Assuming a building in which brittle finishes are to be used, from *Table 5* in the code:

$$\delta_{actual} \le \frac{span}{360} \qquad \therefore \frac{0.104 B.Max\, L^2}{EI} \le \frac{L}{360}$$

where B.Max is the maximum bending moment due to unfactored imposed loads only.

$$\therefore I \ge \frac{37.4 B.Max\, L}{E}$$

**Note:** care must be taken to ensure that a consistent system of units is used. A similar calculation can be carried out for any other $\dfrac{span}{coefficient}$ ratio.

## 2.14  Example 2.8  Deflection of simply supported beam

In Example 2.2 a 305 × 165 × 46 UB was found to be suitable for shear bending web buckling and web bearing. Check the suitability with respect to deflection, assuming brittle finishes to the underside of the beam.

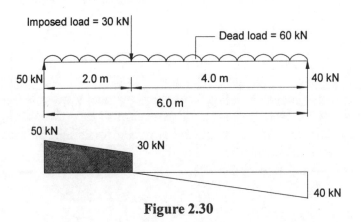

**Figure 2.30**

Maximum bending moment due to unfactored imposed loads $= \dfrac{(50+30)2}{2}$

$$= 80 \text{ kNm}$$

$$\therefore \delta_{\text{actual}} \approx \frac{0.104 B.MaxL^2}{EI} \le \frac{span}{360}$$

*Clause 3.1.2*  Modulus of Elasticity  $E = 205 \text{ kN/mm}^2$
Section property tables  $I_{xx} = 9899 \times 10^4 \text{ mm}^4$

$$\therefore \delta_{\text{actual}} \approx \frac{0.104 \times 80 \times 10^6 \times 6000^2}{205 \times 10^3 \times 9899 \times 10^4} = 14.6 \text{ mm}$$

*Table 5*  $\dfrac{span}{360} = \dfrac{6000}{360} = 16.7 \text{ mm} > \delta_{\text{actual}}$

**Section is adequate with respect to deflection**

This check could have been carried out more accurately using the values given in Table 2.1 of the text.

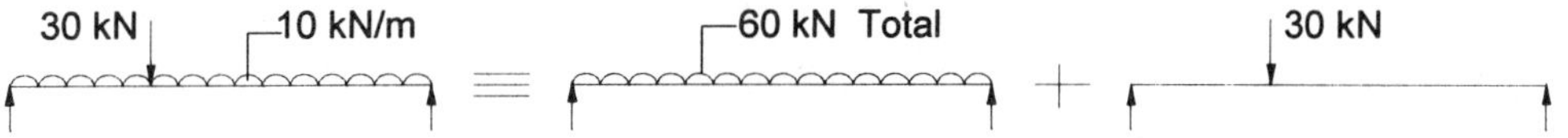

**Figure 2.31**

$$\delta_{\text{actual}} = \frac{5WL^3}{384EI} + \frac{PL^3}{48EI}\left[\frac{3a}{L} - 4\left(\frac{a}{L}\right)^3\right]$$

$$\delta_{\text{actual}} = \frac{5 \times 60 \times 6000^3}{384 \times 205 \times 9899 \times 10^4} + \frac{30 \times 6000^3}{48 \times 205 \times 9899 \times 10^4}\left[\frac{3 \times 2}{6} - 4\left(\frac{2}{6}\right)^3\right]$$

$$\delta_{\text{actual}} = 8.3 + 5.7 = 14 \text{ mm}$$

In this case the approximate technique overestimates the deflection by less than 5%. Provided that the estimated deflection is no more than 95% of the deflection limit, from Table 5 the approximate answer should be adequate for design purposes and a more accurate calculation is not required.

### 2.15  Conservative Method for Lateral Torsional Buckling Moment Capacity (*Clause 4.3.7.7*)

*2.15.1.1  A less rigorous alternative to evaluating the lateral torsional buckling moment capacity than that given in Clause 4.3.7.3 is given in Clause 4.3.7.7. The method applies to equal flanged rolled sections such as universal beams, universal columns and channel sections.*

In this method the buckling resistance moment $M_b$ between lateral restraints is given by:

$$M_b = p_b S_x$$

where:

$p_b$     is determined from *Table 19* for the appropriate $p_y$ value, $\lambda$ and $X$

$\lambda$     is the slenderness of the section taken as $L_E/r$

$L_E$     is the effective length from *Clauses 4.3.5 or 4.3.6* as before,

$r$     is the radius of gyration about the member minor axis, i.e. $r_{yy}$

$X$     is the torsional index as before, i.e. from published section property tables or $D/T$

The slenderness value can be modified to allow for the shape of the bending moment diagram between restraint points by using the correction factor '*n*' from *Table 20*, or more accurately using *Clause 4.3.7.6*, as in the rigorous method demonstrated in section 2.4.4.1 of this text.

### 2.16  Example 2.9  Simply supported beam 1 - conservative method

In Example 2.3 the buckling moment ($M_b$) was determined using the *rigorous method* to be 94.8 kNm. The *conservative method* in determining $M_b$ is as follows:

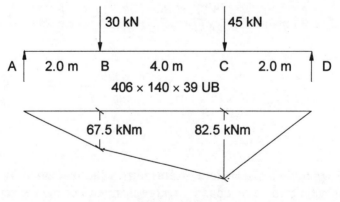

**Figure 2.32**

From section property tables:

   $S_{xx} = 724 \times 10^3$ mm$^3$        $r_{yy} = 28.7$ mm        $X = 47.5$

In this case the bending moment diagram between the restraints at B and C is a combination of two cases given in Table 20.

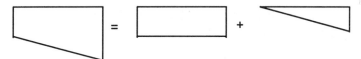

**Figure 2.33**

*Table 20* $\qquad n = 1.0 \qquad\qquad n = 0.77$

An approximate value of '$n$' can be used $= \dfrac{1.0 + 0.77}{2} = 0.89$

effective length $\quad L_E = 0.89 \times 4000 \quad = 3560$ mm

Slenderness $\qquad \lambda = \dfrac{L_E}{r_{yy}} = \dfrac{3560}{28.7} = 124$

*Table 6* $\qquad p_y = 275$ N/mm$^2$ $\ X = 47.5 \qquad \lambda = 124$

*Table 19(a)* $\qquad\qquad\qquad\qquad\qquad\qquad\qquad\qquad$ **$p_b = 119$ N/mm$^2$**

$$M_b = (119 \times 724 \times 10^3)/10^6 = 86.16 \text{ kNm}$$
$$M_b > 82.5 \text{ kNm}$$

**Section is adequate**

It is evident from this method that the value of M$_b$ is underestimated. In situations where a section is proved to be inadequate by a small margin then use of the rigorous method may prove worthwhile in identifying an acceptable economic section.

### 2.17 Example 2.10 Simply supported beam 2 – conservative method

In Example 2.4 the $M_b$ value using the Rigorous Method and grade 50 steel was found to be 197.8 kNm.

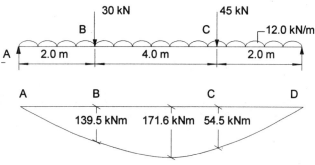

**Figure 2.34**

From section property tables:

$S_{xx} = 1055 \times 10^3$ mm$^3$     $r_{yy} = 38.5$ mm       $X = 38.3$

In this case the bending moment diagram between the restraints at B and C is a combination of two cases given in *Table 20*.

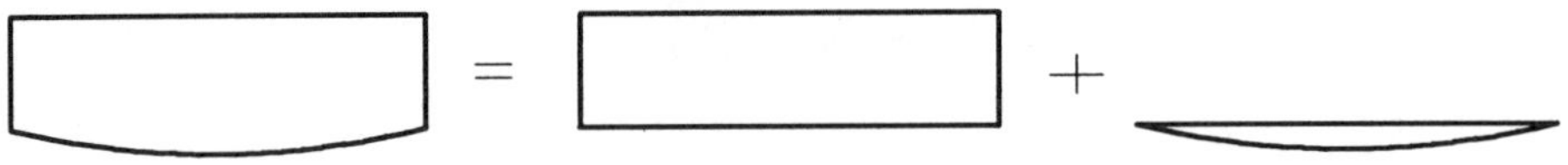

**Figure 2.35**

*Table 20*                              $n = 1.0$                    $n = 0.94$

Average value of   $n \approx \dfrac{1.0 + 0.94}{2} = 0.97$

effective length              $L_E = 0.97 \times 4000 = 3880$ mm

slenderness                   $\lambda = \dfrac{L_E}{r_{yy}} = \dfrac{3880}{38.5} = 101$

*Table 6*        $p_y = 355$ N/mm$^2$    $X = 38.3$           $\lambda = 101$
*Table 19(a)*                                                              $p_b = 177$ **N/mm$^2$**

$M_b = (177 \times 1055 \times 10^3)/10^6 = 187$ kNm
$M_b > 171.6$ kNm

**Section is adequate**

The other checks such as shear, web buckling, web bearing and deflection are all the same as before. The Conservative Method is only applicable to the evaluation of the lateral torsional buckling moment $M_b$.

## 2.18  Safe Load Tables

The Steel Construction Institute publishes member capacity tables to enable a rapid, efficient check on a member subjected to axial load, bending moment or combined axial and bending forces (Ref.:11). The use of such tables to check shear, bending, web buckling and web bearing is illustrated in the following example.

### 2.19 Example 2.11 Beam with intermittent lateral restraint - Use of safe load tables

It is proposed to use a 533 × 210 × 109 UB as a main roof beam spanning 12.0 m and supporting three secondary beams at the mid-span and quarter span points, as shown in Figure 2.36.

- (a) Using the design data and safe load tables given, check the suitability of the section with respect to:

  - (i) shear,
  - (ii) bending,
  - (iii) web buckling     and
  - (iv) web bearing.

- (b) Check the suitability of the beam with respect to deflection assuming brittle finishes.

**Design data:**

| | |
|---|---|
| Characteristic dead load | 5.0 kN/m |
| Characteristic imposed load | 6.0 kN/m |
| Characteristic dead load reaction from each secondary beam | 10.0 kN |
| Characteristic imposed load reaction from each secondary beam | 20.0 kN |

Assume:

- (i) the stiff bearing length for web buckling and web bearing = 25 mm
- (ii) lateral and torsional restraint to the compression flange is provided at the end supports and positions of the secondary beams only.

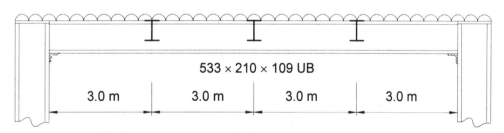

533 × 210 × 109 UB

| 3.0 m | 3.0 m | 3.0 m | 3.0 m |

**Figure 2.36**

**Solution:**

Design distributed load    =    $(1.4 \times 5.0) + (1.6 \times 6.0)$    = 16.6 kN/m

Design beam end reactions    =    $(1.4 \times 10.0) + (1.6 \times 20.0)$    = 46 kN

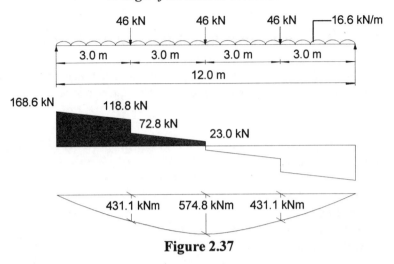

**Figure 2.37**

| Design shear force | = | 168.6 kN |
|---|---|---|
| Design bending moment in portion B-C | = | 574.8 kNm |
| Coincident shear at position of maximum bending moment | = | 23.0 kN |

**Safe Load Table Extracts:**

Table 2.2

| BEARING | | | | | | BUCKLING | | | | | | Shear |
|---|---|---|---|---|---|---|---|---|---|---|---|---|
| End bearing | | | Continous Bearing | | | End bearing | | | Continous Bearing | | | $P_v$ |
| C1 | C2 | C3 | C1 | C2 | C3 | C1 | C2 | C3 | C1 | C2 | C3 | |
| 284 | 3.37 | 8.41 | 568 | 3.37 | 16.8 | 459 | 1.69 | 1.69 | 918 | 1.69 | 3.38 | 1100 |
| **240** | **3.07** | **7.69** | 480 | 3.07 | 15.4 | **371** | **1.38** | **1.38** | 742 | 1.38 | 2.76 | **995** |
| 214 | 2.86 | 7.16 | 428 | 2.86 | 14.3 | 313 | 1.16 | 1.16 | 626 | 1.16 | 2.32 | 922 |

Shear ($P_v$), Web buckling ($P_w$) and Web bearing ($P_{crip}$)

Extract from *Steelwork design guide to BS 5950: Part 1; Volume 1* (The Steel Construction Institute)

Table 2.3

| Buckling Resistance Moment $M_b$(kNm) For Effective Length ($L_e$) in Metres With Variable Slenderness Correction Factors (n) | | | | | | | | | |
|---|---|---|---|---|---|---|---|---|---|
| | n | 1.0 | 1.5 | 2.0 | 2.5 | **3.0** | 3.5 | 4.0 | 4.5 |
| 533 × 210 × 109 | 0.4 | 750 | 750 | 750 | 750 | 750 | 750 | 750 | 750 |
| **Mcx = 750** | 0.6 | 750 | 750 | 750 | 750 | 750 | 735 | 705 | 677 |
| **Plastic** | 0.8 | 750 | 750 | 750 | 739 | 696 | 655 | 614 | 574 |
| | **1.0** | 750 | 750 | 736 | 681 | **625** | 571 | 520 | 473 |

Buckling resistance moment  ($M_b$), Section classification, Moment capacity ($M_c$)

### 2.19.1  Shear

The safe loads in Table 2.2 give the value of $P_v$ as determined from *Clause 4.2.3*

$$P_v = 995 \text{ kN} \gg 168.6 \text{ kN}$$

**Section is adequate in shear**

### 2.19.2  Bending

The moment capacity $M_c$, the section classification and the buckling moment of resistance $M_b$ for a range of effective lengths with various '$n$' values are given in *Table 2.3*. The values of $M_b$ are calculated on the basis of $\lambda_{LT} = nuv\lambda$ as defined in *Clause 4.3.7.5* and in section 2.4.4.1 of this chapter.

| | | |
|---|---|---|
| Section tables | $r_{yy} = 46.0$ mm  effective length = 3000 mm  $N = 0.5$ | |
| Slenderness | $\lambda = \dfrac{L_E}{r_{yy}} = \dfrac{3000}{46.0} = 65.2$ | |
| *Table 18* | $\beta = \dfrac{431.1}{574.8} = 0.75$ | |
| *Table 17* | $M_o = (574.8 - 431.1) = 143.7$ kNm | |
| *Table 16* | $\gamma = \dfrac{M}{M_o} = \dfrac{431.1}{143.7} = 3.0 \qquad n = 0.98 \qquad$ say 1.0 | |

From Extracts of safe load tables   i.e. *Table 2.3* with $n = 1.0$ and $L_E = 3.0$ m

Section is plastic   $M_{cx} = 750$ kNm

$M_b = 625$ kNm $> M_{\text{applied}}$

**Section is adequate in bending**

### 2.19.3  Web Buckling

The capacity for web buckling is given by:

$$P_w = C1 + (b1 \times C2) + (t_p \times C3)$$

where:
C1   is the contribution from the beam,
C2   is the contribution from the stiff bearing length (excluding any additional flange plates),
C3   is the contribution from any continuously welded flange plate if present,
$b1$   stiff bearing length,
$t_p$   thickness of additional flange plate
In this problem there are no additional flange plates present, therefore;

$$P_w = C1 + (b1 \times C2)$$

From Extracts Table 2.2         C1    = 371 kN   and      C2   = 1.38 kN/mm
                                $P_w$   = 371 + (1.38 × 25)      = 405.5 kN
                                $P_w$   >> 168.6 kN

**Section is adequate in web buckling**

### 2.19.4  Web Bearing

Similarly for web bearing
From Extracts Table 2.2           C1 =   240 kN   and   C2 = 3.07 kN/mm
                                $P_{crip}$  =   240 + (3.07 × 25)  = 316.6 kN
                                $P_{crip}$  >> 168.6 kN

**Section is adequate in web bearing**

**Note:**   At the design stage the stiff bearing length required for web buckling and web bearing is often not known, a conservative value of $P_w$ and $P_{crip}$ can be evaluated assuming this to be zero.

   In locations such as the column positions at mid-span, the web buckling and web bearing capacities should be calculated on the basis of the load being distributed in two directions as shown in Figure 2.27, and the continuous buckling/bearing values in Table 2.2 used. In this example the support is the more critical location.

### 2.19.5  Deflection

The deflection is checked for the unfactored imposed load only as shown in Figure 2.38

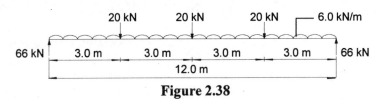

**Figure 2.38**

Maximum bending moment  = (66 × 6) – (20 × 3) × (6 × 6 × 3)  = 228 kNm
Using the approximate method illustrated in Section 2.6 to estimate the deflection
*Clause 3.1.2*    $E = 205$ kN/mm$^2$       From  section tables  $I_{xx}$ =  66820 × 10$^4$  mm$^4$

$$\delta_{actual} \approx \frac{0.104 \times 228 \times 10^6 \times 12000^2}{205 \times 10^3 \times 66820 \times 10^4} = 24.5 \text{ mm}$$

*Table 5*        Limiting $\delta$  = $\dfrac{\text{span}}{360}$ = $\dfrac{12000}{360}$  = 33.3 mm

**Section is adequate in deflection**

A more accurate calculation can be carried out by adding the contribution for each load case, i.e.

$$\delta = \frac{5WL^3}{384EI} + 2\left[\frac{3a}{L} - 4\left(\frac{a}{L}\right)^3\right]\frac{PL^3}{48EI} + \frac{PL^3}{48EI}$$

$$\text{(UDL)} + \left(2 \times \frac{1}{4}\text{ span loads}\right) + \text{(mid-span load)}$$

In this case this calculation is not necessary.

## 2.20 Example 2.12 Pedestrian walkway

Two pedestrian walkways are required for maintenance purposes in the machinery hall of a petro-chemical plant. It is intended to support the walkways using a series of universal beams at 4.5 m centres as shown in Figure 2.39 The proposed surface of the walkways comprises an open grid flooring system which is attached to the top flange of the beams by proprietary brackets. Using the data given, design a suitable universal beam section considering;
- (i)      shear,
- (ii)     bending,
- (iii)    web buckling,
- (iv)    web bearing      and
- (v)     deflection.

**Design data:**

Self-weight of open grid flooring (including hand-rails)          0.1 kN/m

Maximum imposed service load allowed on walkway          3.1 kN/m

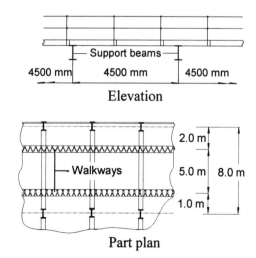

Elevation

Part plan

**Figure 2.39**

Solution to Example 2.12  (see Section 2.22)

## 2.21  Solution to Example 2.2

| Contract : Flexural Members **Job Ref. No. :**<br>**Part of Structure :**   Example 2.2<br>**Calc. Sheet No. :**  1  of  4 | **Calcs. by :** W.McK.<br>**Checked by :**<br>**Date :** |
|---|---|

| References | Calculations | Output |
|---|---|---|
| BS 5950:Part 1 | *Structural use of steelwork in building: Part 1: 1990* | |

*(a)  Design a suitable Universal Beam Section*

Dead load = 10 kN
Imposed load = 30 kN

Dead load = 8 kN/m
Imposed load = 10 kN/m

A ⟶ 2.0 m ⟶ B ⟶ 4.0 m ⟶ C

6.0 m

Design point load  $= (1.4 \times 10) + (1.6 \times 30)$  $=$  62 kN
Design UDL  $= (1.4 \times 8) + (1.6 \times 10)$  $=$  27.2 kN/m

62 kN

27.2 kN/m

2.0 m            4.0 m

$V_A$  ⟶                    ⟶  $V_C$

By proportion:
Vertical reaction at A = $V_A$ =

$$(27.2 \times 3) + \left(\frac{62 \times 4}{6}\right) = 122.9 \text{ kN}$$

Vertical reaction at C = $V_C$ =

$$(27.2 \times 3) + \left(\frac{62 \times 2}{6}\right) = 102.3 \text{ kN}$$

122.9 kN

68.5 kN

6.5 kN

$x$

102.3 kN

Shear Force Diagram

**Output:**

Design shear force
122.9 kN

| Contract : Flexural Members **Job Ref. No. :** | **Calcs. by : W.McK.** |
|---|---|
| **Part of Structure :** Example 2.2 | **Checked by :** |
| **Calc. Sheet No. :** 2 of 4 | **Date :** |

| References | Calculations | Output |
|---|---|---|
| | Position of zero shear $x = \dfrac{6.5}{27.2} = 0.24$ m | |
| | Maximum bending moment occurs at position of zero shear | |
| | B.Max = shaded area = | |
| | $M_x = (122.9 + 68.5) + \left(\dfrac{0.24 \times 6.5}{2}\right) = 192.2$ kNm | Design bending moment   192.2 kNm |
| | Compression flange is fully restrained and assuming low shear | |
| Clause 4.2.5 | Maximum bending moment    $M_x \leq M_c$ | |
| | $\leq p_y S_{xx}$ | |
| | **Note:** It is unlikely that $1.2\,p_y\,Z$ will govern | |
| | $\therefore\ S_{xx} \geq \dfrac{M_x}{p_y}$ | |
| Table 6 | Assume the flange thickness $< 16$ mm $\therefore p_y = 275$ N/mm$^2$ | |
| | $\therefore S_{xx} \geq \dfrac{192.2 \times 10^6}{275} = 698.9 \times 10^3$ mm$^3$ | |
| | A trial beam size can be selected from published section tables    Try a $305 \times 165 \times 46$ UB | |
| Section Tables | $D = 306.6$ mm    $d = 265.2$ mm        $B = 165.7$ mm | |
| | $T = 11.8$ mm      $t = 6.7$ mm        $b/T = 7.02$ | |
| | $d/t = 39.6$ mm  $S_{xx} = 720 \times 10^3$ mm$^3$    $Z_{xx} = 646 \times 10^3$ mm$^3$ | |
| Table 7 | (i)  Section Classification        $\varepsilon = \left(\dfrac{275}{p_y}\right)^{\frac{1}{2}} = 1.0$ | |
| | Flange: Outstand element of compression flange rolled section | |
| | $b/T_{actual} = 7.02\ <\ 7.5\varepsilon$ | Flange is plastic |
| | Web:  Bending only with neutral axis at mid-depth | |
| | $d/t_{actual}\ = 39.6 <\ 79\varepsilon$ | Web is plastic Section is plastic |
| | (ii) Shear | |
| Clause 4.2.3 | $F_v = 122.9$ kN      $P_v = 0.6 p_y A_v = 0.6 p_y t D$ | |
| | $P_v = (0.6 \times 275 \times 6.7 \times 306.6)/10^3 = 338.9$ kN | Section is adequate in shear |
| | $P_v > F_v$ | |

| References | Calculations | Output |
|---|---|---|
| | **Contract :** Flexural Members **Job Ref. No. :**<br>**Part of Structure :** Example 2.2<br>**Calc. Sheet No. :** 3 of 4 | **Calcs. by :** W.McK.<br>**Checked by :**<br>**Date :** |

| References | Calculations | Output |
|---|---|---|
| *Clause 4.2.5* | (iii) Bending<br>$60\% \, P_v = 0.6 \times 338.9 = 203.3$ kN $> 68.5$ kN | Low shear |
| | Shape factor $= v = \dfrac{720}{646} = 1.11 \; < \; 1.2 \quad \therefore \; p_y S_{xx}$ governs | |
| | $\qquad\qquad S_{xx \, \text{provided}} > S_{xx \, \text{required}}$ | Section is adequate in bending |
| | (b) Design a suitable Rectangular Hollow Section<br>Try a $300 \times 200 \times 8$ RHS | |
| *Section Tables* | $A = 7710 \; \text{mm}^2 \quad D = 300 \text{ mm} \quad B = 200 \text{ mm}$<br>$Z_{xx} = 653 \times 10^3 \; \text{mm}^3 \qquad\quad S_{xx} = 785 \times 10^3 \; \text{mm}^3$<br>$d/t = 34.5 \qquad\qquad\qquad b/t = 22.0$ | |
| *Table 6* | $t < 16.0$ mm $\quad p_y = 275$ N/mm$^2$ $\qquad$ **Note:** $\varepsilon = 1.0$ | |
| *Table 7* | (i) Section Classification $\quad \varepsilon = \left(\dfrac{275}{p_y}\right)^{\frac{1}{2}} = 1.0$ | |
| | Flange: Internal element of compression flange<br>$\qquad\qquad$ rolled section<br>$\qquad\qquad b/t_{\text{actual}} = 22 \; < \; 26\varepsilon$<br>Web: $\quad$ Bending only with neutral axis at mid-depth<br>$\qquad\qquad d/t_{\text{actual}} \;\; = 34.5 < \; 79\varepsilon$ | Flange is plastic<br><br>Web is plastic<br>Section is plastic |
| *Clause 4.2.3* | (ii) Shear<br>$F_v = 122.9$ kN $\qquad\qquad P_v = 0.6 p_y A_v = 0.6 p_y \left(\dfrac{D}{D+B}\right) A$ | |
| | $P_v = \left( 0.6 \times 275 \times \left(\dfrac{300}{300+200}\right) 7710 \right) \Big/ 10^3 = 763.3$ kN | |
| | $\qquad\qquad\qquad P_v \gg F_v$ | Section is adequate in shear |
| *Clause 4.2.5* | (iii) Bending<br>$60\% \, P_v = 0.6 \times 763.3 = 458$ kN $> 68.5$ kN $>$ (coincident shear) | Low shear |
| | Shape factor $= v = \dfrac{785}{653} = 1.2 \quad$ The 1.2 factor may be | |
| | replaced by the average load factor | |
| | average load factor $\approx \dfrac{\text{working load}}{\text{factored load}}$ | |
| | factored loads $\quad = \quad 62 + (6 \times 27.2) \quad = \quad 225.2$ kN<br>working loads $\quad = \quad 40 + (6 \times 18) \quad\;\; = \quad 148.0$ kN | |

| | | |
|---|---|---|
| **Contract :** Flexural Members **Job Ref. No. :** | | **Calcs. by :** W.McK. |
| **Part of Structure :** Example 2.2 | | **Checked by :** |
| **Calc. Sheet No. :** 4 of 4 | | **Date :** |

| References | Calculations | Output |
|---|---|---|
| | average load factor $\approx \dfrac{225.2}{148.0} = 1.52$ <br><br> $M_c = \quad p_y\,S_{xx} \quad \leq \quad 1.52 p_y\,Z_{xx}$ <br> $p_y\,S_{xx} \qquad = \quad (275 \times 785 \times 10^3)\,/\,10^6 \qquad = \quad 215.9 \text{ kNm}$ <br> $1.52 p_y\,Z_{xx} \qquad = (1.52 \times 275 \times 653 \times 10^3)\,/\,10^6 = 273 \text{ kNm}$ <br><br> $\qquad\qquad\qquad M_c = 215.9 \text{ kNm} > M_x$ | Section is adequate in bending |

## 2.22  Solution to Example 2.12

| | | |
|---|---|---|
| **Contract :** Flexural Members  **Job Ref. No. :**<br>**Part of Structure :**  Example 2.12<br>**Calc. Sheet No. :**  1  **of**  4 | | **Calcs. by :** W.M$^c$K.<br>**Checked by :**<br>**Date :** |

| References | Calculations | Output |
|---|---|---|
| *BS 5950:Part 1* | *Structural use of steelwork in building: Part 1: 1990*<br><br>Design load due to each walkway =<br>                  =  [(1.4 × 0.1) + (1.6 × 3.1)]4.5<br>                  =  23 kN<br>Assume the self-weight of the beam equal to 1.4 kN/m<br>  Design load due to self-weight = 1.4 × 1.4 ≈ 2.0 kN/m<br><br><br><br>Design shear force         = 33.9 kN<br>Design bending moment  = 52.5 kNm<br>Coincident shear force    = 24.1 kN<br>**Note:**  The coincident shear force used is adjacent to the<br>       position of maximum bending moment<br><br>   Try a 457 × 152 × 52 UB, this is selected on the basis of<br>   experience or the use of safe load tables. | |
| *Section Property*<br>*Tables* | $D = 449.8$ mm     $d = 407.6$ mm   $I_{xx} = 21370$ cm$^4$<br>$B = 152.4$ mm     $b/T = 6.99$      $Z_{xx} = 950$ cm$^3$<br>$t = 7.6$ mm        $d/t = 53.6$      $S_{xx} = 1096$ cm$^3$<br>$T = 10.9$ mm     $r_{yy} = 3.11$ cm  $u = 0.859$  $X = 43.9$ | |
| *Table 6*<br>*Table 7* | $T < 16$           $p_y = 275$ N/mm$^2$<br>   Outstand of compression flange    $b/T = 6.99 < 8.5\varepsilon$<br>Web with neatral axis at mid-depth  $d/t = 53.6 < 79\varepsilon$ | Flange is plastic<br>Web is plastic<br>**Section is plastic** |

| Contract : Flexural Members **Job Ref. No. :** <br> Part of Structure : Example 2.12 <br> Calc. Sheet No. : 2 of 4 | Calcs. by : W.M$^c$K. <br> Checked by : <br> Date : |
|---|---|

| References | Calculations | Output |
|---|---|---|
| *Clause 4.2.3* | Design shear force $F_v$ = 33.9 kN <br> Shear capacity $\quad P_v = [0.6 \times 275 \times (7.6 \times 449.8)]/10^3$ <br> $\qquad\qquad\qquad = 564$ kN <br> $\qquad\qquad\qquad P_v \gg F_v$ | **Adequate in shear** |
| *Clause 4.2.5* | Coincident shear $\quad = 24.1$ kN $\ll 0.6\,P_v$ <br><br> Shape factor $\quad v = \dfrac{S_{xx}}{Z_{xx}} = \dfrac{1096}{950} \quad = 1.15 \quad <1.2$ <br> $\therefore \ M_c = p_y S_{xx} = (275 \times 1096 \times 10^3)/10^6 = 301.4$ kNm | |
| *Clause 4.3.7.1* | $\overline{M} \le M_b \quad$ where $\qquad \overline{M} = mM_A$ <br> The compression flange is restrained at the ends only | |
| *Clause 4.3.7.6* <br> *Table 13* <br> *Clause 4.3.7.3* <br> *Clause 4.3.74* | $m \ = 1.0 \qquad\qquad \overline{M} = 52.5$ kNm <br><br> $M_b \ = \ S_{xx}\,p_b$ <br> $\lambda_{LT} = \ nuv\lambda$ | |
| *Clause 4.3.7.5* | $\gamma \ = \ \dfrac{M}{M_o} \qquad M = 0 \qquad \therefore \gamma = 0$ | |
| *Tables 13,16* <br> Section Tables | $\beta \ = \ 0$ | $n = 0.94$ <br> $u = 0.859$ |
| *Table 14* | $N = 0.5\,\lambda \ = \ \dfrac{8000}{31.1} = 257 \qquad \dfrac{\lambda}{X} = \dfrac{257}{43.9} = 5.9$ <br> $\lambda_{LT} = \ (0.94 \times 0.859 \times 0.77 \times 257) \qquad = 160$ | $v = 0.77$ <br><br> $\lambda_{LT} = 160$ |
| *Table 11* | $\lambda_{LT} = 160, \qquad p_y = 275$ N/mm$^2$ $\qquad\qquad p_b \ = 60$ N/mm$^2$ <br><br> $M_b = (1096 \times 10^3 \times 60\ )/10^6 = 65.8$ kNm <br> $M_b > \overline{M}$ | **Adequate in bending** |
| *Clause 4.5.2* | Web buckling $\qquad P_w = (n_1 + b_1)t\,p_c$ <br><br> Assume $b_1 = 0 \quad n_1 \ = \ \dfrac{D}{2} \ = \dfrac{449.8}{2} = 224.9$ <br><br> $t = 7.6$ mm $\qquad \lambda \ = 2.5\dfrac{d}{t} = \ 2.5 \times 53.6 \ = 134.0$ | |
| *Table 27(c)* | $p_c = 82$ N/mm$^2$ <br> $P_w = \ (224.9 \times 7.6 \times 82)/10^3 = 140$ kN <br> $P_w > \ 33.9$ kN | **Adequate in web buckling** |

| References | Calculations | Output |
|---|---|---|
| | | |

**Contract :** Flexural Members  **Job Ref. No. :**
**Part of Structure :**   Example 2.12
**Calc. Sheet No. :**  3  **of**  4

**Calcs. by :** W.M$^c$K.
**Checked by :**
**Date :**

---

*Clause 4.5.3*

Web bearing          $P_{crip} = (n_1 + b_2)tp_{yw}$

$$b_2 = 2.5 \times \left(\frac{D-d}{2}\right) = 2.5 \times \left(\frac{449.8 - 407.6}{2}\right) = 21.1$$

$P_{crip} = (21.1 \times 7.6 \times 275)/10^3 = 44.1$ kN

$P_{crip} > 33.9$ kN

**Adequate in web bearing**

*Clause 2.5*

Assuming no brittle finishes    $\delta \leq \dfrac{span}{200} = \dfrac{8000}{200} = 40$ mm

$$\delta_{actual} \approx \frac{0.104 B. Max L^2}{EI}$$

Unfactored imposed load due to each walkway
$$= 3.1 \times 4.5 = 14.0 \text{ kN}$$

```
        14.0 kN              14.0 kN
          ↓                    ↓
     ┌── 2.0 m ──┬─── 5.0 m ───┬ 1.0 m ┐
12.25 kN                              15.75 kN
```

```
     12.25 kN
  ████████████
             └──────────────┐
        1.75 kN             │
                            15.75 kN
```

Maximum bending moment = $(12.25 \times 2.0) = 24.5$ kNm

$$\delta \approx \frac{0.104 \times 24.5 \times 10^6 \times 8000^2}{205 \times 10^3 \times 21370 \times 10^4} = 3.7 \text{ mm}$$

$$\delta_{actual} \ll \delta_{limited}$$

**Adequate in deflection**

Alternatively, a more accurate calculation for deflection can be determined from:

$$\delta_{actual} = \frac{PL^3}{48EI}\left[\frac{3a}{L} - 4\left(\frac{a}{L}\right)^3\right] \quad + \quad \frac{PL^3}{48EI}\left[\frac{3a}{L} - 4\left(\frac{a}{L}\right)^3\right]$$

          due to left hand load               due to right-hand load

| Contract : Flexural Members **Job Ref. No. :** | Calcs. by : W.M$^c$K. |
|---|---|
| Part of Structure : Example 2-12 | Checked by : |
| Calc. Sheet No. : 4 of 4 | Date : |

| References | Calculations | Output |
|---|---|---|
| | $$\delta_{\text{actual}} = \frac{14 \times 8^3 \times 10^8}{48 \times 205 \times 10^3 \times 21370} \left[ \frac{3 \times 2}{8} - 4\left(\frac{2}{8}\right)^3 \right] \quad +$$ $$\frac{14 \times 8^3 \times 10^8}{48 \times 205 \times 10^3 \times 21370} \left[ \frac{3 \times 1}{8} - 4\left(\frac{1}{8}\right)^3 \right] = 3.6 \text{ mm}$$ | |

# 3 Axially Loaded Members

## 3.1 Introduction

The design of axially loaded members considers any member where the applied loading induces either axial tension or axial compression. Members subject to axial forces most frequently occur in bracing systems, pin-jointed trusses, lattice girders or suspension systems, as shown in Figure 3.1

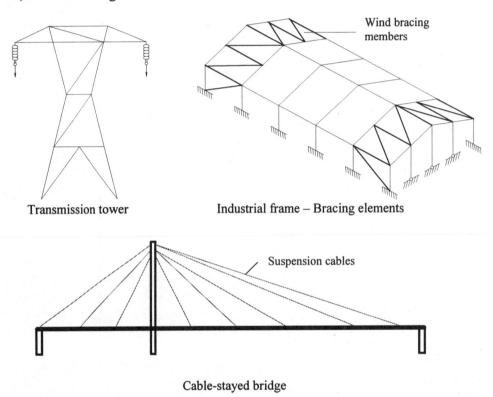

Transmission tower      Industrial frame – Bracing elements

Cable-stayed bridge

**Figure 3.1**

Frequently, in structural frames, sections are subjected to combined axial and bending effects which may be caused by eccentric connections, wind loading or rigid-frame action. In most cases in which UB and UC sections are used as columns in buildings, they are subjected to combined axial and bending effects. The design of such members is discussed and illustrated in Chapter 4.

The types of section used for axially loaded members range from rolled uniform beams, columns and hollow sections to threaded bars, flat plates and wire ropes.

The following discussion relates primarily to pin-jointed structures which comprise the majority of structures with members subject to axial loads only.

## 3.2    Pin-jointed Frames

The use of beams and plate girders as discussed in Chapters 2 and 6 does not always provide the most economic or suitable structural solution when spanning large openings. In buildings which have lightly loaded long span roofs, when large voids are required within the depth of roof structures for services, when plate girders are impractical, or for aesthetic/architectural reasons, the use of roof trusses or lattice girders may be more appropriate.

Trusses are frequently used as secondary structural elements to distribute wind loading to the foundations, as temporary bracing during construction and for torsional and lateral stability.

Roof trusses and lattice girders are open-web flexural members which transmit the effects of loads applied within their spans to support points by means of bending and shear. In the case of beams, the bending and shear is transmitted by inducing bending moments and shear forces in the cross-sections of structural members. Trusses and lattice girders, however, generally transfer their loads by inducing axial tensile or compressive forces in the individual members. The form of a truss is most economic when the arrangement is such that most members are in tension.

The magnitude and sense of these forces can be determined using standard methods of analysis such as *the method of sections*, *joint resolution*, *tension coefficients*, *graphical techniques* or the use of *computer software*.

The arrangement of the internal framing of a roof truss depends upon its span. Rafters are normally divided into equal panel lengths and ideally the loads are applied at the node points by roof purlins. Purlin spacing is dependent on the form of roof cladding that is used and is usually based on manufacturers' data sheets. In instances where the purlins do not coincide with the node points, the main members (i.e. the rafters, or the top and bottom booms of lattice girders) are also subjected to local bending which must be allowed for in the design.

The internal structure of trusses should be such that, where possible, the long members are ties (in tension), while the short members are struts (in compression). In long span trusses the main ties are usually cambered to offset the visual sagging effects of the deflection.

In very long span trusses, e.g. 60 metres, it is not usually possible to maintain a constant slope in the rafter owing to problems such as additional heating requirements caused by the very high ridge height. This problem can be overcome by changing the slope to provide a *mansard-type* truss in which the slope near the end of the truss is very steep, while it is shallower over the rest of the span.

A few examples of typical pitched roof trusses are illustrated in Figure 3.2. Lattice girders are generally trusses with parallel top and bottom chords (known as booms) with internal web bracing members. In long span construction they are very useful since their relatively small span/depth ratio (typically 1/10 to 1/14) gives them an advantage over pitched roof trusses.

The two most common types are the Pratt-Truss (N-Girder) and the Warren Truss, both of which are shown in Figure 3.3. The top chord often has a camber to assist drainage from the roof.

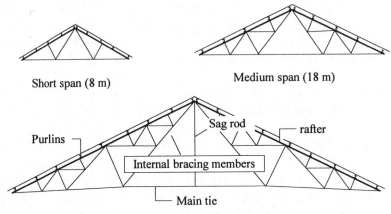

Short span (8 m)

Medium span (18 m)

Large span (30 m)

**Figure 3.2**

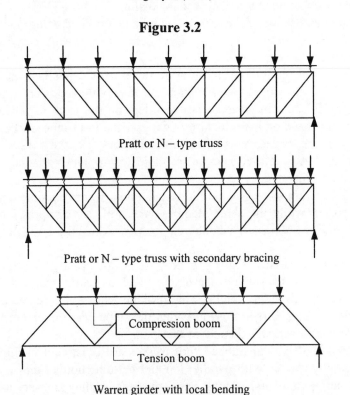

Pratt or N – type truss

Pratt or N – type truss with secondary bracing

Warren girder with local bending

**Figure 3.3**

As with roof trusses, the framing should be triangulated, considering the span and the spacings of the applied loads. If purlins do not coincide with the panel points then secondary bracing as shown in Figure 3.3 can be adopted as an alternative to designing for combined axial and local bending effects (see Example 3.6).

Generally, the four main assumptions made when analysing trusses are:

♦ *Truss members are connected together at their ends only.*

In practice, the top and bottom chords are normally continuous and span several joints rather than being a series of discontinuous, short members. Since truss members are usually long and slender and do not support significant bending moments this assumption in the analysis is acceptable.

The design of such members is carried out on the basis of combined axial and approximate bending moment effects, as illustrated in Chapter 4, Example 4.4 and given in *Clause 4.8* of BS 5950:Part 1.

♦ *Truss members are connected together by frictionless pins*

In real trusses the members are connected at the joints using bolted or welded gusset plates or end plates as shown in Figure 3.4, rather than frictionless pins.

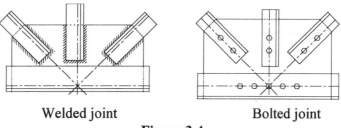

Welded joint                Bolted joint

**Figure 3.4**

Provided that the setting-out lines of the bolts or the centroidal axes of the members intersect at the assumed joint locations, experience has shown that this idealisation is acceptable.

♦ *The truss structure is loaded only at the joints.*

Often the exact location of purlins relative to the joints on the top of the compression chord/rafters is unknown at the design stage of a truss. In these circumstances, assuming that the purlins do not coincide with the position of the joints, a local bending moment in addition to the axial load is assumed in the truss members.

A number of empirical design rules are given in *Clause 4.10* for this situation. In addition, *Clause 4.10* indicates that secondary stresses induced by the inherent stiffness of the assumed pin-joints between the members, may be ignored provided that:

(i) the slenderness of the chord members in the plane of the truss is greater than 50 and

(ii) the slenderness of most of the web members is greater than 100.

♦ *The self-weight of the members may be neglected or assumed to act at the adjacent nodes.*

Frequently, in the analysis of small trusses, it is reasonable to neglect the self-weight of the members. This may not be acceptable for large trusses, particularly those used in bridge construction. Common practice is to assume that half of the weight of each member acts at each of the two joints that it connects.

## 3.3 Résumé of Analysis Techniques

The following résumé gives a brief summary of the most common manual techniques adopted to determine the forces induced in the members of statically determinate pin-jointed frames. There are numerous structural analysis books available which give comprehensive detailed explanations of these techniques.

### 3.3.1 Method of Sections

The *method of sections* involves the application of the three equations of static equilibrium to two-dimensional plane frames, i.e.

| | | | | |
|---|---|---|---|---|
| +ve ↑ | $\Sigma F_y$ | $= 0$ | the sum of the vertical forces equals zero | equation (i) |
| +ve → | $\Sigma F_x$ | $= 0$ | the sum of the horizontal forces equals zero | equation (ii) |
| +ve ↗ | $\Sigma M$ | $= 0$ | the sum of the moments of the forces taken about anywhere in the plane of the frame equals zero | equation (iii) |

The sign convention indicated in equations (i) (ii) and (iii) for positive directions has been adopted in this text. The sign convention adopted to indicate ties and struts in frames is as shown in Figure 3.5

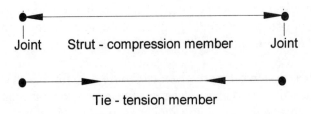

**Figure 3.5**

The method involves considering an imaginary section line which cuts the frame under consideration into two parts A and B, as shown in Figure 3.8.

Since only three independent equations are available any section taken through a frame must not include more than three members for which the internal force is unknown.

Consideration of the equilibrium of the resulting force system enables the magnitude and sense of the forces in the cut members to be determined.

## 3.4 Example 3.1 Pin-jointed truss

A pin-jointed truss simply supported by a pinned support at A and a roller support at J carries three loads at nodes C, D and E as shown in Figure 3.6. Determine the magnitude and sense of the forces induced in members X, Y and Z as indicated.

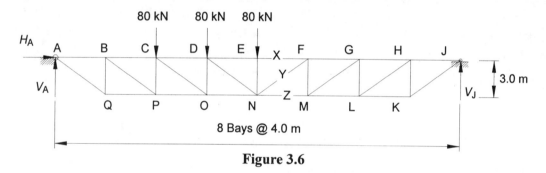

**Figure 3.6**

**Step 1:** evaluate the support reactions. It is not necessary to know any information regarding the frame members at this stage other than dimensions as shown in Figure 3.7 since only externally applied loads and reactions are involved.

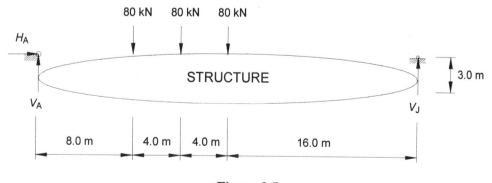

**Figure 3.7**

Apply equations (i), (ii) and (iii) to the force system

+ve ↑    $\Sigma F_y = 0$    $V_A - (80 + 80 + 80) + V_J = 0$   ·   $V_A + V_J = 240$ kN

+ve →    $\Sigma F_x = 0$                                                    $H_A = 0$

+ve $\nearrow$    $\Sigma M_A = 0$    $(80 \times 8) + (80 \times 12) + (80 \times 16) - (V_J \times 32)$        $= 0$

$$V_J = \textbf{90 kN}$$

hence    $V_A = \textbf{150 kN}$

The moment equation can be considered about **any** point in the plane of the frame. Point "A" is used since the line of action of both $H_A$ and $V_A$ passes through this point and consequently these unknowns do not appear in the resulting equation.

**Step 2:** select a section through which the frame can be considered to be cut and using the same three equations of equilibrium determine the magnitude and sense of the unknown forces (i.e. the internal forces in the cut members).

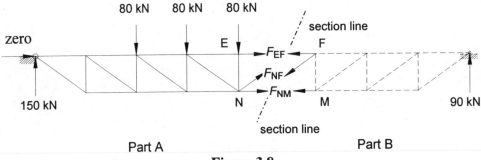

**Figure 3.8**

It is convenient to **assume** all unknown forces to be tensile and hence at the cut section their direction and lines of action are considered to be pointing away from the joint (refer to Figure 3.8). If the answer results in a negative force, this means that the assumption of a tie was incorrect and the member is actually in compression, i.e. a strut.

The application of the equations of equilibrium to either part of the cut frame will enable the forces X, Y and Z to be evaluated.

**Note:** the section considered must not cut through more than three members with unknown internal forces.

Consider Part A

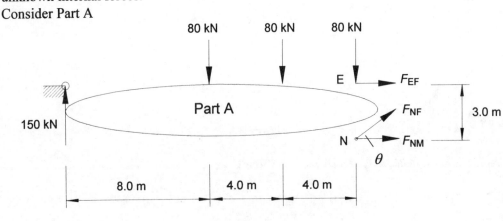

**Figure 3.9**

**Note:** $\sin \theta = \dfrac{3}{5} = 0.6,$ $\cos \theta = \dfrac{4}{5} = 0.8$

+ve ↑ $\quad \Sigma F_y = 0 \quad 150 - (80 + 80 + 80) + F_{NF}\sin \theta = 0$

$$F_{NF} = \frac{90}{\sin \theta} = \frac{90}{0.6} = +150 \text{ kN}$$

**Member NF is a TIE**

+ve → $\quad \Sigma F_x = 0 \quad F_{EF} + F_{NM} + F_{NF}\cos \theta = 0$

+ve ⟋ $\quad \Sigma M_N = 0 \quad (150 \times 16) - (80 \times 8) - (80 \times 4) + (F_{EF} \times 3) = 0$

$$F_{EF} = -480 \text{ kN}$$

**Member EF is a STRUT**

hence $F_{NM} = -F_{EF} - F_{NF}\cos \theta = -(-480) - (150 \times 0.8) = +360 \text{ kN}$

**Member NM is a TIE**

These answers can be confirmed by considering Part B of the structure and applying the equations as above.

### 3.4.1   Joint Resolution

Considering the same frame using joint resolution highlights the advantage of the method of sections when only a few member forces are required.

In this technique, (which can be considered as a special case of the method of sections), sections are taken which isolate each individual joint in turn in the frame, e.g.

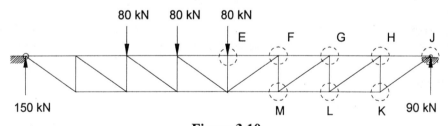

**Figure 3.10**

In Figure 3.10 eight sections are shown, each of which isolates a joint in the structure as indicated in Figures 3.11(a) and (b).

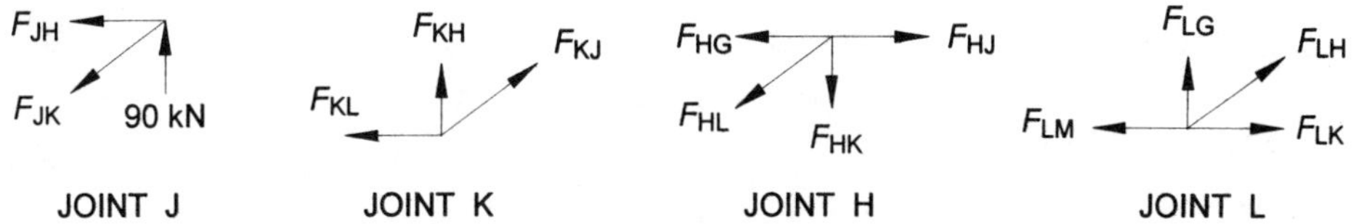

**Figure 3.11(a)**

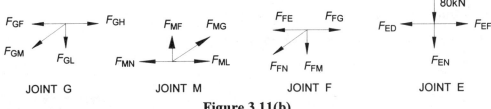

**JOINT G**          **JOINT M**          **JOINT F**          **JOINT E**

**Figure 3.11(b)**

Since in each case the forces are coincident, the moment equation is of no value; hence only two independent equations are available. It is necessary when considering the equilibrium of each joint to do so in a sequence which ensures that there are no more than two unknown member forces in the joint under consideration. This can be carried out until all member forces in the structure have been determined.

Consider Joint J:

$+ve \uparrow \Sigma F_y = 0 \qquad +90 - F_{JK}\cos\beta \qquad\qquad = 0$

$$F_{JK} = +\frac{90}{0.6} \qquad\qquad = +150 \text{ kN}$$

$+ve \rightarrow \Sigma F_x = 0 \qquad -F_{JH} - F_{JK}\sin\beta \qquad = 0$

$\cos\beta = 0.6 \qquad\qquad\qquad F_{JH} = -(150 \times 0.8) \qquad = -120 \text{ kN}$

$\sin\beta = 0.8 \qquad$ **Member JK is a TIE and member JH is a STRUT**

Consider Joint K:

$+ve \uparrow \Sigma F_y = 0 \qquad +F_{KH} + 150\sin\theta \qquad = 0$

$\qquad\qquad\qquad\qquad\qquad F_{KH} = -(150 \times 0.6) \qquad = -90 \text{ kN}$

$+ve \rightarrow \Sigma F_x = 0 \qquad -F_{KL} + 150\cos\theta \qquad = 0$

$\cos\beta = 0.8$

$\sin\beta = 0.6 \qquad\qquad\qquad F_{KL} = -(150 \times 0.8) \qquad = +120 \text{ kN}$

**Member KH is a STRUT and member KL is a TIE**

Consider Joint H:

$+ve \uparrow \Sigma F_y = 0 \qquad +90 - F_{HL}\cos\beta \qquad\qquad = 0$

$$F_{HL} = +\frac{90}{0.6} \qquad\qquad = +150 \text{ kN}$$

$+ve \rightarrow \Sigma F_x = 0 \qquad -120 - F_{HG} - F_{HL}\sin\beta \qquad = 0$

$\cos\beta = 0.6 \qquad\qquad\qquad F_{HG} = -(150 \times 0.8) \qquad = -20 \text{ kN}$

$\sin\beta = 0.8 \qquad$ **Member HL is a TIE and member HG is a STRUT**

Consider Joint L:

$+ve \uparrow \Sigma F_y = 0 \qquad +F_{LG} + 150\sin\theta \qquad\qquad = 0$

$\qquad\qquad\qquad\qquad\qquad F_{LG} = -(150 \times 0.6) \qquad = -90 \text{ kN}$

$+ve \rightarrow \Sigma F_x = 0 \qquad -F_{LM} + 150\cos\theta + 120 \qquad = 0$

$\qquad\qquad\qquad\qquad\qquad F_{LM} = -(150 \times 0.8) + 120 \quad = +240 \text{ kN}$

$\sin\theta = 0.6$
$\cos\theta = 0.8$

**Member LG is a STRUT and member LM is a TIE**

Consider Joint G:

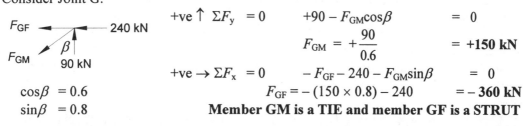

$\cos\beta = 0.6$
$\sin\beta = 0.8$

$$+ve \uparrow \ \Sigma F_y \ = 0 \qquad +90 - F_{GM}\cos\beta \qquad = 0$$
$$F_{GM} = +\frac{90}{0.6} \qquad = +150 \text{ kN}$$
$$+ve \rightarrow \Sigma F_x \ = 0 \qquad -F_{GF} - 240 - F_{GM}\sin\beta \qquad = 0$$
$$F_{GF} = -(150 \times 0.8) - 240 \qquad = -360 \text{ kN}$$

**Member GM is a TIE and member GF is a STRUT**

Consider Joint M:

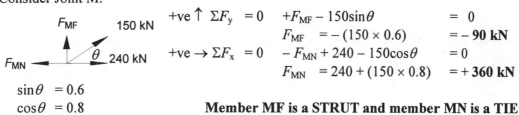

$\sin\theta = 0.6$
$\cos\theta = 0.8$

$$+ve \uparrow \ \Sigma F_y \ = 0 \qquad +F_{MF} - 150\sin\theta \qquad = 0$$
$$F_{MF} = -(150 \times 0.6) \qquad = -90 \text{ kN}$$
$$+ve \rightarrow \Sigma F_x \ = 0 \qquad -F_{MN} + 240 - 150\cos\theta \qquad = 0$$
$$F_{MN} = 240 + (150 \times 0.8) \qquad = +360 \text{ kN}$$

**Member MF is a STRUT and member MN is a TIE**

Consider Joint F:

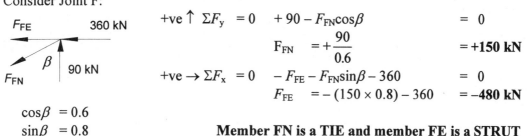

$\cos\beta = 0.6$
$\sin\beta = 0.8$

$$+ve \uparrow \ \Sigma F_y \ = 0 \qquad +90 - F_{FN}\cos\beta \qquad = 0$$
$$F_{FN} = +\frac{90}{0.6} \qquad = +150 \text{ kN}$$
$$+ve \rightarrow \Sigma F_x \ = 0 \qquad -F_{FE} - F_{FN}\sin\beta - 360 \qquad = 0$$
$$F_{FE} = -(150 \times 0.8) - 360 \qquad = -480 \text{ kN}$$

**Member FN is a TIE and member FE is a STRUT**

Consider Joint E:

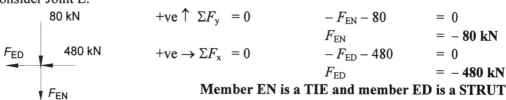

$$+ve \uparrow \ \Sigma F_y \ = 0 \qquad -F_{EN} - 80 \qquad = 0$$
$$F_{EN} \qquad = -80 \text{ kN}$$
$$+ve \rightarrow \Sigma F_x \ = 0 \qquad -F_{ED} - 480 \qquad = 0$$
$$F_{ED} \qquad = -480 \text{ kN}$$

**Member EN is a TIE and member ED is a STRUT**

Graphical techniques have largely been superseded by the use of computer software for all but the most basic trusses. In addition, tables of coefficients are available which enable the rapid evaluation of member forces for a variety of structural forms, e.g. *Timber Designers' Manual* [Ref:20].

### *3.4.2    Tension Coefficients*

The method of tension coefficients is a tabular technique of carrying out joint resolution in either two or three dimensions. It is ideally suited to the analysis of pin-jointed spaceframes.

Consider an individual member from a pin-jointed plane-frame, e.g. member AB shown in Figure 3.12 with reference to a particular X-Y co-ordinate system.

If AB is a member of length $L_{AB}$ having a tensile force in it of $T_{AB}$, then the components of this force in the X and Y directions are $T_{AB} \cos\theta$ and $T_{AB} \sin\theta$ respectively.

If the co-ordinates of A and B are $(X_A, Y_A)$ and $(X_B, Y_B)$, then the component of $T_{AB}$ in the x-direction is given by :

$$\text{x-component} \quad = \quad T_{AB}\frac{(X_B - X_A)}{L_{AB}} = \quad t_{AB}(X_B - X_A)$$

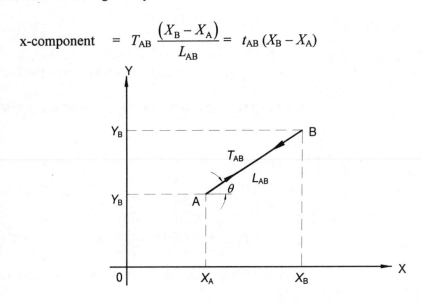

**Figure 3.12**

where $t_{AB} = \dfrac{T_{AB}}{L_{AB}}$ and is known as the **tension coefficient** of the bar. Similarly, the component of $T_{AB}$ in the y-direction is given by :

$$\text{y-component} \quad = \quad T_{AB}\frac{(Y_B - Y_A)}{L_{AB}} \quad = \quad t_{AB}(Y_B - Y_A)$$

If at the joint A in the frame there are a number of bars, i.e. AB, AC ... AN, and external loads $X_A$ and $Y_A$ acting in the X and Y directions, then since the joint is in equilibrium the sum of the components of the external and internal forces must equal zero in each of those directions.

Expressing these conditions in terms of the components of each of the forces then gives:

$$t_{AB}(X_B - X_A) + t_{AC}(X_C - X_A) + \ldots\ldots\ldots\ldots \quad t_{AN}(X_N - X_A) + X_A = 0 \qquad (1)$$

$$t_{AB}(Y_B - Y_A) + t_{AC}(Y_C - Y_A) + \ldots\ldots\ldots\ldots .t_{AN}(Y_N - Y_A) + Y_A = 0 \qquad (2)$$

A similar pair of equations can be developed for each joint in the frame giving (2 × No. of joints) equations in total. In a triangulated frame the number of unknown forces (i.e. members) is equal to [(2 × No. of joints) −3], hence there are three additional equations which can be used to determine the reactions or check the values of the tension coefficients.

Once a tension coefficient (e.g. $t_{AB}$) has been determined, the unknown member force is given by the product

$$T_{AB} = t_{AB}L_{AB}$$

**Note:** A member which has a −ve tension coefficient is in compression and is therefore a strut.

## 3.5 Example 3.2 Two dimensional plane-truss

Consider the pin-jointed, plane-frame ABC loaded as shown in Figure 3.13.

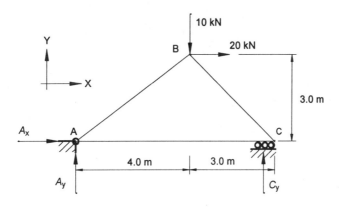

**Figure 3.13**

Construct a table in terms of tension coefficients and an X/Y co-ordinate system as shown in Table 3.1.

The equilibrium equations are solved in terms of the '$t$' values and hence the member forces and support reactions are evaluated and entered in the table as shown in Table 3.1.

**Table 3.1**

| Joint | | Equilibrium Equations | Member | $t$ | Length (m) | Force (kN) |
|---|---|---|---|---|---|---|
| **A** | X | $4t_{AB} + 7t_{AC} + A_x = 0$ | AB | *1.43* | 5 | *+7.15* |
| | Y | $3t_{AB} \qquad + A_y = 0$ | AC | ? | 7 | ? |
| | | | BC | *–4.76* | $3\sqrt{2}$ | *–20.19* |
| **B** | X | $-4t_{AB} + 3t_{BC} + 20 = 0$ | **Support Reactions** | | | |
| | Y | $-3t_{AB} - 3t_{BC} - 10 = 0$ | **Support** | **Component** | | **Reaction** |
| **C** | X | $-7t_{AC} - 3t_{BC} \qquad = 0$ | **A** | $A_x$ | | ? |
| | Y | $+3t_{BC} + C_y \qquad = 0$ | | $A_y$ | | ? |
| | | | **C** | $C_y$ | | ? |

Consider joint B:
There are only two unknowns and two equations, hence:
Adding both equations

$$
\begin{aligned}
-4t_{AB} + 3t_{BC} + 20 &= 0 \\
-3t_{AB} - 3t_{BC} - 10 &= 0 \\
\hline
-7t_{AB} \qquad\quad + 10 &= 0 \qquad\quad t_{AB} = 1.43
\end{aligned}
$$

substitute for $t_{AB}$ in the first equation $\qquad\qquad t_{BC} = -4.76$

Force in member $\qquad$ AB $= t_{AB} \times L_{AB} = (1.43 \times 5.0) \qquad = +7.15$ kN $\qquad$ **TIE**

Force in member $\qquad$ BC $= t_{BC} \times L_{BC} = \left(-4.76 \times 3\sqrt{2}\right) = -20.19$ kN $\qquad$ **STRUT**

Joints A and C can be considered in a similar manner until all unknown values, including reactions, have been determined. The reader should complete this solution.

In the case of a space frame, each joint has three co-ordinates and the forces have components in the three orthogonal X, Y and Z directions. This leads to (3 × No. of joints) equations which can be solved as above to determine the '$t$' values and subsequently the member forces and support reactions.

## 3.6   Example 3.3   Three dimensional space truss

The space frame shown in Figure 3.14 has three pinned supports at A, B and C, all of which lie on the same level as indicated. Member DE is horizontal and at a height of 10 m above the plane of the supports. Determine the forces in the members when the frame carries loads of 80 kN and 40 kN acting in a horizontal plane at joints E and D respectively (see Table 3.2)

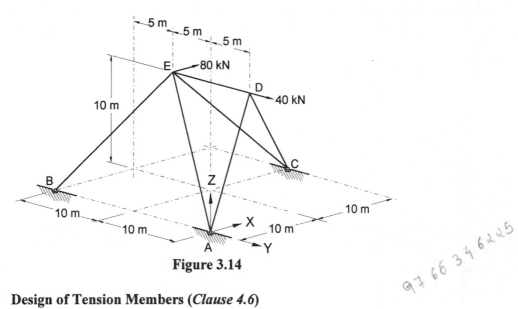

**Figure 3.14**

## 3.7   3.7  Design of Tension Members (*Clause 4.6*)

As indicated in Section 3.1, the types of element used in the design of tension members is numerous and varied; open and closed single-rolled sections being adopted for light trusses and lattice girders, compound sections comprising either multiple-rolled sections or welded plates for heavy trusses with ropes and cables being used in suspension structures such as bridges and roofs.

There are a number of potential problems that may arise from using light, slender sections such as bars, flats, rolled angle and channel sections, e.g.

- ◆ excessive sag under self-weight,
- ◆ vibration during dynamic loading        and
- ◆ damage during transportation to site.

The introduction of sag rods as indicated in Figure 3.2 and the use of intermediate packing in double angle or channel members will assist in minimizing the first two of these problems.

In general, if the leg length of an angle tie is at least equal to $\dfrac{\text{member length}}{60}$, the member will have sufficient stiffness to prevent damage during transport.

The tension capacity $P_t$ of a member should be determined from

$$P_t = A_e \times p_y$$

where:

$p_y$ is defined in *Table 6*     and
$A_e$ is the effective area of the section as defined in *Clauses 3.3.3 or 4.6.2 to 4.6.4*

**Table 3.2**

| Joint | | Equilibrium Equations | Member | $t$ | Length (m) | Force (kN) |
|---|---|---|---|---|---|---|
| **A** | X | $+10t_{AE} + 10t_{AD} + A_X = 0$ | AE | $0$ | 20.6 | $0$ |
| | Y | $-15t_{AE} - 5t_{AD} + A_Y = 0$ | AD | $0$ | 15 | $0$ |
| | Z | $+10t_{AE} + 10t_{AD} + A_Z = 0$ | DC | $0$ | 15 | $0$ |
| **B** | X | | DE | $+4.0$ | 10 | $+40.0$ |
| | Y | | EC | $-4.0$ | 15 | $-60.0$ |
| | Z | | EB | $+4.0$ | 15 | $+60.0$ |
| **C** | X | | **Support Reactions** | | | |
| | Y | | **Support** | **Component** | | **Reaction** |
| | Z | | A | $A_x$ | | $0$ |
| **D** | X | | | $A_y$ | | $0$ |
| | Y | | | $A_z$ | | $0$ |
| | Z | | B | $B_x$ | | $-40.0$ |
| **E** | X | | | $B_y$ | | $-20.0$ |
| | Y | | | $B_z$ | | $-40.0$ |
| | Z | | C | $C_x$ | | $-40.0$ |
| Readers should complete and solve the equilibrium equations and hence confirm the member forces and support reactions indicated in the right-hand column | | | | $C_y$ | | $-20.0$ |
| | | | | $C_z$ | | $+40.0$ |
| Tip: Identify a joint with no more than three unknowns to begin solving the equations. | | | | | | |

*Clause 3.3.3*
Experimental testing has indicated that the effective capacity of a member in tension is not reduced by the presence of holes provided that the ratio:

$$\frac{\text{net cross - sectional area}}{\text{gross cross - sectional area}} > \frac{\text{yield strength}}{\text{ultimate strength}} \quad \text{by a suitable margin.}$$

This margin is reflected in the differing $K_e$ values adopted for different grades of steel, as indicated in Table 3.3.

**Table 3.3**

| Grade of steel | 40 , 43 | 50 , WR50 | 55 | All others |
|---|---|---|---|---|
| $K_e$ | 1.2 | 1.1 | 1.0 | $0.75\dfrac{U_s}{Y_s} \le 1.2$ |

where:
$U_s$   is the specified minimum ultimate tensile strength
$Y_s$   is the specified minimum yield strength

The effective area $A_e = K_e \times$ net area
where the net area is defined in *Clause 3.3.2* as the gross-area less deductions for fastener holes as given in *Clause 3.4*.

*Clause 3.4:* **Deductions for holes**
*Clause 3.4.1:*
When deducting the area of a fastener hole from a cross-section, allowance should be made for the clearance dimensions, e.g. a 20 mm diameter bolt requires a 22 mm diameter hole (see Chapter 5, section 5.2).

*Clause 3.4.2:* **Holes not staggered**
If holes are not staggered across the width of a member, then the area to be deducted is the maximum sum of the sectional areas of the holes in any cross-section perpendicular to the direction of the applied stress in the member.
For example, consider the flat plate tie member shown in Figure 3.15.

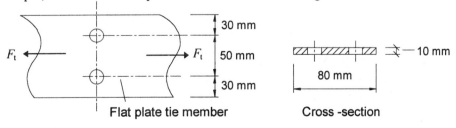

**Figure 3.15**

hole diameter            =  20 + 2        =  22 mm
Area to be deducted      =  (22 × 10)2    =  440 mm$^2$

### Clause 3.4.3  Staggered holes

If holes are staggered across the width of a member then it is necessary to consider all possible failure paths extending progressively across the member. The area to be deducted is equal to the sum of the sectional areas of all holes in the path less an allowance equal to $\dfrac{s_p^2 t}{4g}$ for each gauge space in the path as shown in Figures 3.16 and 3.17.

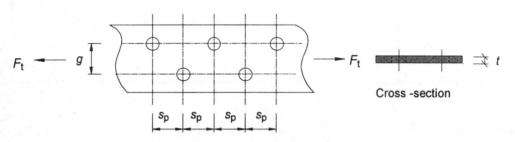

**Figure 3.16**

where:

$s_p$     is the staggered pitch
$g$       is the gauge
$t$       is the thickness

For example, consider the flat plate tie member shown in Figure 3.17.

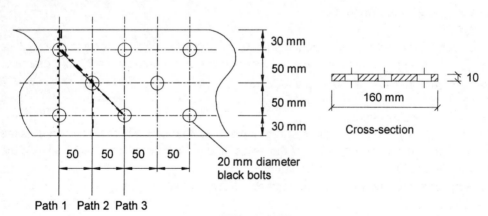

**Figure 3.17**

Path 1     Area to be deducted     =  (22 × 10)2                           =  440 mm$^2$

Path 2     Area to be deducted     =  $(22 \times 10)2 - \left( \dfrac{50^2 \times 10}{4 \times 50} \right)$

                                    =  440 – 125                           =  315 mm$^2$

Path 3    Area to be deducted    $= (22 \times 10)3 - \left(\dfrac{50^2 \times 10}{4 \times 50}\right)2$

$= 660 - 250$                    $= 410 \text{ mm}^2$

Path 1 is the most severe with the highest area to be deducted and therefore:

Net area                 $A_{net} = (160 \times 10) - 440$                    $= 1160 \text{ mm}^2$

Assuming grade 43 steel is being used

$A_e = K_e \times A_{net}$    ($\le$ gross area)  $= 1600 \text{ mm}^2$

Effective area    $A_e = 1.2 \times 1160$                    $= 1392 \text{ mm}^2$

### *Clause 4.6.2* **Effects of eccentric connections**

Although theoretically, tension members are inherently stable and the most economic structural elements, the introduction of secondary effects such bending due to eccentricities at connections reduces their efficiency. With the exception of angles, channels and T-sections, the secondary effects should be allowed for in the design by considering members subject to combined axial and bending load effects (see Chapter 4).

### *Clause 4.6.3.1* **Single angles, Channels and T-Sections**

For asymmetric connections where secondary effects will occur, the effective area should be calculated as indicated in Figure 3.18.

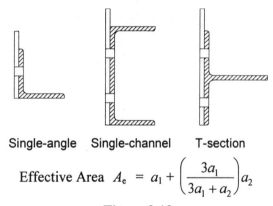

Single-angle   Single-channel   T-section

Effective Area  $A_e = a_1 + \left(\dfrac{3a_1}{3a_1 + a_2}\right)a_2$

**Figure 3.18**

where :

$a_1$    is the net sectional area of the connected leg

$a_2$    is the sectional area of the unconnected leg

**Note:** the area of a leg of an angle is defined in *Clause 4.6.3.2* as '*...the product of the thickness by the length from the outer corner minus half the thickness, and the area of the leg of a T-section is the product of the thickness by the depth minus the thickness of the flange*', as shown in Figure 3.19.

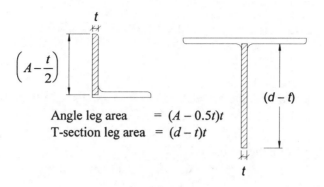

Angle leg area     $= (A - 0.5t)t$
T-section leg area $= (d - t)t$

**Figure 3.19**

### *Clause 4.6.3.2* **Double angles**

For double angles connected back-to-back and to the same side of a gusset plate or section as shown in Figure 3.20, a similar calculation is made to determine the effective area.

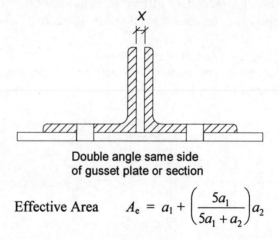

Double angle same side
of gusset plate or section

Effective Area     $A_e = a_1 + \left( \dfrac{5a_1}{5a_1 + a_2} \right) a_2$

**Figure 3.20**

where $a_1$ and $a_2$ are as before.

In double angle ties where $x$ > aggregate thickness of the legs connected with solid packing pieces or the slenderness of an individual angle > 180, each angle should be regarded as a single angle and the effective area calculated as in *Clause 4.6.3.1*.

### *Clause 4.6.3.3*

In symmetrical connections such as two angles back-to back or members connected using lug angles as shown in Figure 3.21, the effective area should be calculated as for flat plates using *Clauses 3.3.2 to 3.4*.

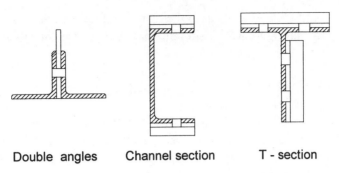

Double angles    Channel section    T - section

**Figure 3.21**

### Example 3.4  Lattice girder

Consider the lattice girder in Example 3.1 and Figure 3.6, and check the suitability of the sections shown in Table 3.4 for members NM and JK. The applied loading is assumed to be the factored design loading and 16 mm diameter black bolts are to be used throughout.

**Table 3.4**

| Member | Axial Load | Section |
|---|---|---|
| Bottom chord member NM | + 360 kN | 2/ 75 × 50 × 8 Angles, long legs back-to-back connected with gusset plate between the angles |
| Diagonal tie member JK | + 150 kN | 1/ 75 × 50 × 8 Angle, long leg connected to gusset plate |

Member NM:   2/ 75 × 50 × 8   Double angles

*Clause 4.6.1*          $P_t = A_e p_y$
*Table 6*               $T < 16$ mm          $p_y = 275$ N/mm$^2$

*Clause 4.6.3.3(b)*   and   *Clause 3.3.2*
$$A_e = A_{net} = (A_{gross} - \text{area of holes})$$

Section property tables   $A_{gross} = 18.9 \times 10^2$ mm$^2$
Area of hole = $(18 \times 8)2$          $= 288$ mm$^2$
$A_e = (1890 - 288)$          $= 1602$ mm$^2$
$P_t = (1602 \times 275)/10^3$   $= 441$ kN      $> 360$ kN

**Section is adequate**

The two angles should be held apart by bolted packing pieces in at least two places along their length, as shown in Figure 3.22, using the same bolt size as the end connection.

### Clause 4.6.3.3
The outermost packing piece should be approximately (9 × smallest leg length) from the end connection, i.e. $(9 \times 50) = 450$ mm.

Although *Clause 4.6.3.2* specifically states that in double angles connected back to back to one side of a gusset or section, the slenderness of the individual components should not exceed 80; this is usually also applied to double angles connected to both sides of a gusset or section, therefore:

$$\text{slenderness } \lambda \ = \ \frac{spacing\ of\ packing\ pieces}{r_{vv}} \ \le \ 80$$

$\therefore$ spacing of packing pieces $\le 80 \times 10.7 = 856$ mm

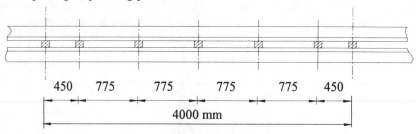

| 450 | 775 | 775 | 775 | 775 | 450 |

4000 mm

**Figure 3.22**

Member JK: 1/ 75 × 50 × 8 Angle

*Clause 4.6.3.1*    Effective Area    $A_e = a_1 + \left(\dfrac{3a_1}{3a_1 + a_2}\right)a_2$

$$a_1 = (75 - 4)8 - (18 \times 8) \qquad = 424 \text{ mm}^2$$
$$a_2 = (50 - 4)8 \qquad\qquad\quad = 368 \text{ mm}^2$$
$$A_e = 424 + \left(\frac{3 \times 424}{3 \times 424 + 368}\right)368 \ = 709 \text{ mm}^2$$
$$P_t = (709 \times 275)/10^3 \qquad = 195 \text{ kN} \ > \ 150 \text{ kN}$$

**Section is adequate**

### Design of Compression Members (*Clause 4.7*)

The design of compression members is more complex than that of tension members and encompasses the design of structural elements referred to as columns, stanchions or struts. The term strut is usually used when referring to members in lattice/truss frameworks, while the other two generally refer to vertical or inclined members supporting floors and/or roofs in structural frames.

As with tension members, in many cases such members are subjected to both axial and bending effects. This chapter deals primarily with those members on lattice/truss frameworks in which it is assumed that all members are subjected to concentric axial loading. Column/stanchion design in which combined axial compression and bending are present is discussed in Chapter 4.

The dominant mode of failure to be considered when designing struts is axial buckling. Buckling failure is caused by secondary bending effects induced by factors such as:

- ♦ the inherent eccentricity of applied loads due to asymmetric connection details,
- ♦ imperfections present in the cross-section and/or profile of a member throughout its length,
- ♦ non-uniformity of material properties throughout a member.

The first two of these factors are the most significant and their effect is to introduce initial curvature, secondary bending and consequently premature failure by buckling before the stress in the material reaches the yield value. The stress at which failure will occur, known as the *compressive strength* ($p_c$), is influenced by several variables, e.g.

- ♦ the cross-sectional shape of the member,
- ♦ the slenderness of the member,
- ♦ the yield strength of the material,
- ♦ the pattern of residual stresses induced by differential cooling of the member after the rolling process.

The effects of these variables are reflected in *Tables 25, 26* and *27(a), (b), (c)* and *(d)* in the code, which are used to determine the appropriate value of $p_c$ for a particular circumstance.

*A practical and realistic assessment of the critical slenderness of a strut is the most important criterion in determining the compressive strength.*

### 3.9.1 Slenderness    (Clause 4.7.3)

Slenderness is evaluated using:

$$\lambda = \frac{L_e}{r}$$

where:
$\lambda$    is the slenderness ratio,
$L_e$    is the effective length with respect to the axis of buckling being considered,
$r$    is the radius of gyration with respect to the axis of buckling being considered.

Limiting values of slenderness are given in the code to reduce the possibility of premature failure in long struts, and to ensure an acceptable degree of robustness in a member. These limits, which are given in *Clause 4.7.3.2* are shown in Table 3.5 of this chapter.

**Table 3.5**

| Type of member | $\lambda_{maximum}$ |
|---|---|
| Members resisting dead and/or imposed load | 180 |
| Members resisting wind loads only in addition to self-weight | 250 |
| Members which normally act as ties but are subject to compression due to stress reversal | 350 |

In addition to the values in Table 3.5, members in which the slenderness is greater than 180 should be checked for self-weight deflection. In cases where this deflection is greater than (0.001 × length), then the secondary bending effects should be considered in the design.

*Effective Lengths*   (*Clause 4.7.2*)

The effective length is considered to be the actual length of the member between points of restraint multiplied by a coefficient to allow for effects such as stiffening due to end connections of the frame of which the member is a part. Appropriate values for the coefficients are given in *Table 24* of the code and illustrated in Figure 3.23(a) and (b).

In the case of angles, channels and T-sections, secondary bending effects induced by end connections can be ignored and pure axial loading assumed, provided that the slenderness values are determined using *Clauses 4.7.10.2* to *4.7.10.5* or *Table 28* in the code. The use of these *Clauses* and *Table 28* is illustrated in Examples 3.5 and 3.6.

In the case of other cross-sections the slenderness should be evaluated using effective lengths as indicated in Figures 3.23(a) and (b). In addition, Appendix D of the code gives the appropriate coefficients to be used when assessing the effective lengths for columns in single-storey buildings using simple construction; this is dealt with in Chapter 4.

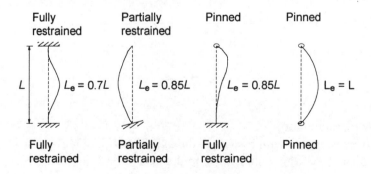

**Figure 3.23(a)**

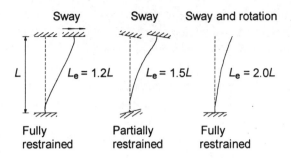

**Figure 3.23(b)**

The coefficients given for determining effective lengths are generally greater than those predicted by mathematical theory; this is to allow for effects such as the inability in practice to obtain full fixity.

### 3.9.2  Compressive Resistance     (*Clause 4.7.4*)

In *Clause 4.7.4*, the compressive resistance $P_c$ of a member is given by:

$$P_c = A_g p_{cs} \quad \text{for slender sections} \quad \text{and}$$
$$P_c = A_g p_c \quad \text{for all other sections}$$

where:
$A_g$    is the gross cross-sectional area as defined in *Clause 3.3.1*
$p_{cs}$   is the compressive strength for slender sections as defined in *Clause 3.6* and *4.7.5*
$p_c$    is the compressive strength as defined in *Clause 4.7.5*

#### 3.9.2.1  Gross Cross-sectional Area   (*Clause 3.3.1*)

The gross cross-sectional area of a member is determined using the member profile and size ignoring any holes required for fasteners, but allowing for any additional holes. When considering the design of laced, battened struts or splices, the cross-section of the laces, battens and splices should be ignored.

#### 3.9.2.2  Slender Sections     (*Clause 3.6.4*)

Where an element has been defined as slender using the criteria from *Table 7*, and it is subject to compression, the design strength $p_y$ should be modified by a factor which is given in *Table 8* of the code for various cross-sections. This reduced value of $p_y$ is then used when extracting values of $p_c$ (compressive strength) from *Tables 27(a)* to *(d)*.

### 3.9.2.3  *Compressive Strength*  (*Clause 4.7.5*)

When determining the value of $p_c$ for a section, reference is first made to *Table 25*. If additional flange plates have been welded to rolled I or H-sections, or if welded plate I or H-sections are used then further reference is made to *Table 26*. This is necessary since the residual stresses induced by welding differ in both magnitude and distribution from those caused during the rolling/cooling processes and effectively further reduce the stress at which buckling may occur. *Table 25* indicates the relevant *Table 27 (a)*, *(b)*, *(c)* or *(d)* to use in determining the appropriate $p_c$ value and relates to the type of cross-section, thickness of elements and axis of buckling being considered.

*Table 27* requires the $p_y$ value and slenderness of the section to determine the tabulated $p_c$ value. When using *Table 27* to determine the $p_c$ value for sections fabricated by welded plates, the $p_y$ value should be reduced by 20 N/mm$^2$.

Compound struts comprising two components back-to-back and connected symmetrically to both sides of a gusset are less stable about an axis through the plane of the connection of the components than an equivalent solid section. The slenderness is therefore checked using an additional ratio $\lambda_b$ as given in *Clauses 4.7.9* and *4.7.13*.

$$\lambda_b = \left(\lambda_m^2 + \lambda_c^2\right)^{\frac{1}{2}}$$

where:

$\lambda_m$     is the ratio of the effective length to the radius of gyration of a whole member about the axis perpendicular to the plane of the connection,

$\lambda_c$     is the ratio of the effective length of a main component to its minimum radius of gyration.

**Note:** The main components should be connected at intervals such that the member is divided into a minimum of three bays of approximately equal length.

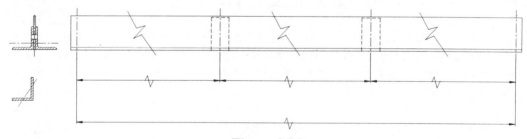

**Figure 3.24**

### 3.9.2.4  *Example 3.5 Pin-Jointed Truss*

Consider the lattice girder in Example 3.1 and Figure 3.6 and check the suitability of the sections shown in Table 3.5 for members FM and EF. The applied loading is assumed to be factored design loading and 16 mm diameter black-bolts are to be used.

**Table 3.5**

| Member | Axial Load | Section |
|---|---|---|
| Vertical strut member FM | – 90 kN | 1/100×75×10 angle with long leg connected to 12 mm thick gusset plates and double bolted at the ends. |
| Top chord member EF | – 480 kN | 2/150×75×15 angles with long legs back-to-back, double-bolted to 12 mm thick gusset plates and with packing pieces single-bolted at the quarter length positions |

**Member FM:** 1/100×75×10 angle long leg double-bolted to gusset plate.

*Table 6*   thickness < 16 mm   $\therefore p_y = 275$ N/mm$^2$
*Table 7*   $b = 75$ mm   $d = 100$ mm   $T = 10$ mm

$$\varepsilon = 1.0 \qquad \frac{b}{T} = \frac{75}{10} = 7.5 \quad < \quad 8.5\varepsilon$$

$$\frac{d}{t} = \frac{100}{10} = 10.0 \quad < \quad 9.5\varepsilon$$

$$\frac{b+d}{T} = \frac{175}{10} = 17.5 \quad < \quad 23\varepsilon$$

**Section is semi-compact**

*Clause 4.7.2* To determine effective length and slenderness use either *Table 24* or *Clause 4.7.10*. In structures such as the one considered here, it is normal to refer to *Clause 4.7.10* for discontinuous strut members and in the case of continuous members i.e. the compression boom of a lattice girder, as an alternative to refer to *Table 24*.

*Clause 4.7.10.2.(a)* and *Table 28*.
For single-angle struts double-bolted by one leg to a gusset plate or other member:

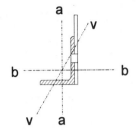

slenderness    $\geq$   $0.85L_{vv}/r_{vv}$
                $\geq$   $0.7L_{vv}/r_{vv} + 15$
                $\geq$   $1.0L_{aa}/r_{aa}$
                $\geq$   $0.7L_{aa}/r_{aa} + 30$
                $\geq$   $0.85L_{bb}/r_{bb}$
                $\geq$   $0.7L_{bb}/r_{bb} + 30$

where:

$L_{aa}$, $L_{bb}$ and $L_{vv}$ are taken as the length '$L$' between the intersections of the centroidal axes or the setting-out lines of the bolts.

**Note:** *Clauses 4.7.10.3 to 4.7.10.4* give similar criteria for double-angles struts, single channel struts and single T-section struts respectively.

*Clause 4.7.3*  Limiting Slenderness

Assuming that the loading is combined dead and imposed     $\lambda \leq 180$

$L_{aa} = L_{bb} = L_{vv}$  =  3000 mm    $r_{aa} = 21.6$    $r_{bb} = 31.2$ mm      $r_{vv} = 15.9$ mm

$$
\begin{aligned}
0.85L_{vv}/r_{vv} &= (0.85 \times 3000)/15.9 &&= 160 \\
0.7L_{vv}/r_{vv} + 15 &= (0.7 \times 3000)/15.9 + 15 &&= 147 \\
1.0L_{aa}/r_{aa} &= (1.0 \times 3000)/21.6 &&= 139 \\
0.7L_{aa}/r_{aa} + 30 &= (0.7 \times 3000)/21.6 + 30 &&= 127 \\
0.85L_{bb}/r_{bb} &= (0.85 \times 3000)/31.2 &&= 81.7 \\
0.7L_{bb}/r_{bb} + 30 &= (0.7 \times 3000)/31.2 + 30 &&= 97.3
\end{aligned}
$$

**The critical slenderness is 160 < 180      $\therefore$  acceptable**

*Clause 4.7.4 (a)*                    $P_c = A_g p_c$
*Clause 3.3.1*            Gross area      $A_g = 1660$ mm$^2$
*Table 25*     Buckling about any axis use *Table 27(c)*

| $\lambda$ $\diagdown$ $p_y$ | 265 | 275 | 305 |
|---|---|---|---|
| 155 | 64 | 65 | 66 |
| **160** | 61 | 61 | 63 |
| 165 | 58 | 58 | 60 |

**Extract from *Table 27(c)* BS 5950:Part 1**

From the extract of *Table 27(c)*      $p_c$  = 61 N/mm$^2$
                $P_c$  = $(1660 \times 61)/10^3 = 101$ kN    > 90 kN

**Section is adequate**

**Member EF**:   2/150×75×15 double angles, long leg connected to gusset plate.

*Table 6*    thickness < 16 mm                    ∴ $p_y = 275$ N/mm$^2$
*Table 7*    $b = 75$ mm        $d = 150$ mm        $T = 15$ mm

$$\varepsilon = 1.0 \qquad \frac{b}{T} = \frac{75}{15} \ = \ 5.5 \ \ < \ 8.5\varepsilon$$

$$\frac{d}{t} = \frac{150}{15} \ = \ 10.0 \ \ < \ 9.5\varepsilon$$

$$\frac{b+d}{T} = \frac{225}{15} \ = \ 15 \ \ < \ 23\varepsilon$$

**Section is semi-compact**

*Clause 4.7.10.2.(c)* and *Table 28*

slenderness $\geq \ 0.85 L_{xx}/r_{xx}$

$\geq \ 0.7 L_{xx}/r_{xx} + 30$

$\geq \ \left[ \left( L_{yy}/r_{yy} \right)^2 + \lambda_c^2 \right]^{\frac{1}{2}}$

$\geq \ 1.4\lambda_c$

where:

$L_{xx}$ and $L_{yy}$ are taken as the length '$L$' between the intersections of the centroidal axes or the setting out lines of the bolts.

$\lambda_c = \dfrac{L_{vv}}{r_{vv}} \leq 50 \qquad$ and

$L_{vv}$ is the distance between intermediate packing pieces
$r_{vv}$ is the radius of gyration for a single angle about the v-v axis
$L_{xx} = L_{yy} \ = \ 4000$ mm        $r_{xx} = 47.5$    $r_{yy} = 31.0$ mm        $r_{vv} = 15.8$ mm

For a 150×75×15 double angle since $\lambda_c \leq 50$  $L_{vv} \leq \ 50 \times 15.8 = \ 790$ mm

Assume the member is divided into 6 bays    ∴ $L_{vv} = \dfrac{4000}{6} = \ 667$ mm

$$\lambda_c = \frac{667}{15.8} = 42.19$$

$\lambda \geq \ 0.85 L_{xx}/r_{xx} \qquad = \ (0.85 \times 4000)/47.5 \qquad = \ 71.6$

$\geq \ 0.7 L_{xx}/r_{xx} + 30 \qquad = \ (0.7 \times 4000)/47.5 + 30 \quad = \ 88.9$

$\geq \ \left[ \left( L_{yy}/r_{yy} \right)^2 + \lambda_c^2 \right]^{\frac{1}{2}} = \ \left[ \left( 4000/31.0 \right)^2 + 42.19^2 \right]^{\frac{1}{2}} = \ 135.8$

$\geq \ 1.4\lambda_c \qquad\qquad = \ 1.4 \times 42.19 \qquad\qquad = \ 59.1$

**The critical slenderness is 135.8 < 180        ∴ acceptable**

*Clause 4.7.4 (a)*                    $P_c = A_g p_c$
*Clause 3.3.1*        Gross area      $A_g = 6330 \text{ mm}^2$
*Table 25* Buckling about any axis use *Table 27(c)*

| $\lambda$ $\diagdown$ $p_y$ | 265 | 275 | 305 |
|---|---|---|---|
| 130 | 85 | 86 | 89 |
| 135 | 80 | 81 | 84 |
| 140 | 76 | 76 | 79 |

**Extract from *Table 27(c)* BS 5950:Part 1**

From the extract of *Table 27(c)*        $p_c = 80.2 \text{ N/mm}^2$
$$P_c = (6330 \times 80.2)/10^3 = 508 \text{ kN} > 480 \text{ kN}$$

**Section is adequate**

### 3.10  Example 3.6  Secondary bracing in lattice girder

A rugby club requires a new stand and, while full uninterrupted views would be desirable, the cost of a cantilever roof is prohibitive. An alternative solution is where a series of castellated beams are simply supported on a series of rectangular hollow sections at the rear and four Pratt trusses at the front, as shown in Figures 3.25 (a), (b) and (c). Using the design data given, check the suitability of the proposed section size for the members of the trusses.

**Note:** neglect wind loading

**Design Data:**
Characteristic dead load due to self weight of roof decking etc.          1.0 kN/m$^2$
Characteristic imposed load                                              0.6 kN/m$^2$
Top and bottom chords and diagonals                      200 × 120 × 8.0 RHS

As mentioned in Section 3.2, when loads are applied to the chords of lattice girders between the node points, a system of secondary bracing can be used to transfer the forces to the node points and main girder members.

In this example each of the four main Pratt trusses supports a series of beams some of which are located at the mid-span points between the nodes in the top chord. The loads applied to the chord by these beams can be transmitted through a secondary bracing system using additional members, as shown in Figure 3.26. It is convenient when analysing this structural arrangement to consider the truss as the superposition of two systems as shown.

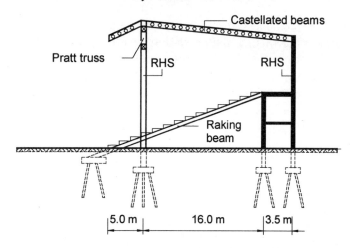

**Figure 3.25 (a)** Typical internal frame

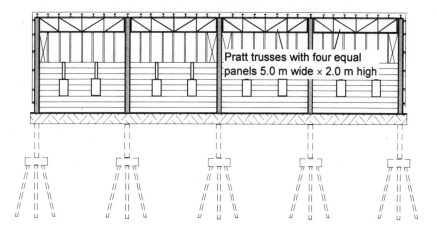

**Figure 3.25 (b)** Front elevation of Grandstand

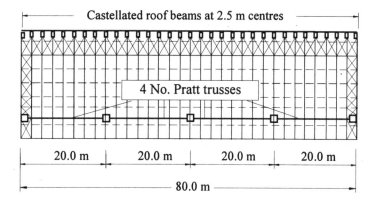

**Figure 3.25 (c)** Roof plan

When determining the member forces for design purposes, the 'mid-span' loads are distributed to adjacent nodes using simple statics, resulting in a Primary load system as shown in Figure 3.26. A standard pin-jointed, plane-frame analysis is carried out to determine the magnitude and sense of the Primary loads.

The secondary bracing elements in combination with some of the main elements transfer the loads induced in the assumed static distribution, as shown in Figure 3.26. The magnitude and sense of the secondary forces can be determined as before.

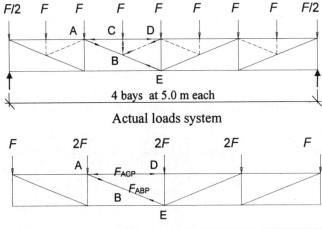

Actual loads system

Primary load system

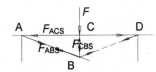

Secondary load system

**Figure 3.26**

The total force in any member is determined from the sum of the Primary and the Secondary forces, for example;

Force in member AB $= F_{abp} + F_{abs}$
Force in member AC $= F_{acp} + F_{acs}$
Force in member CB $= \phantom{F_{acp} +} F_{cbs}$

**Note:** there is no Primary component for member CB.

**Solution:**
Area supported by each castellated beam $\approx$ $2.5 \times (5 + 16 + 3.5)$ $= 61.25\text{m}^2$
Design load/m$^2$ $= (1.4 \times 1.0) + (1.6 \times 0.6)$ $= 2.36 \text{ kN/m}^2$
Total design load /beam $= 2.36 \times 61.25$ $= 144.55 \text{ kN}$
Each castellated beam is simply supported with a 5.0 m cantilever overhang.

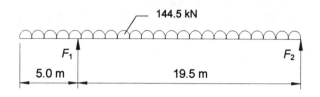

**Figure 3.27**

The support reaction $F_1$ for the beam equals the magnitude of the point load applied to the Pratt truss.

$\sum$Moments about RH end $\qquad$ $(F_1 \times 19.5) - (144.5 \times 12.25) = 0$

$$F_1 = 90.81 \text{ kN}$$
$$2F_1 = 181.6 \text{ kN}$$

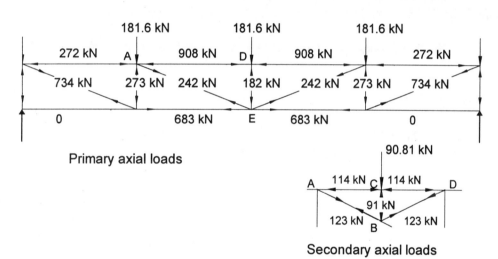

Primary axial loads

Secondary axial loads

Equivalent Load System

**Figure 3.28**

| | | | | |
|---|---|---|---|---|
| Force in member AC | $F_{AC}$ | $= -(908 + 114)$ | $= -1022$ kN | strut |
| Force in member AB | $F_{AB}$ | $= +(242 + 123)$ | $= +365$ kN | tie |
| Force in member BD | $F_{BD}$ | $= +(0 + 123)$ | $= +123$ kN | tie |
| Force in member BC | $F_{BC}$ | $= -(273 + 91)$ | $= -364$ kN | strut |

Similarly for all other member forces.

Top chord member AB: 200×120×8.0 RHS
Applied axial load $\quad = \quad 1002$ kN compression
*Table 6* $\qquad t \ = \ 6.3$ mm $\ < \ 16.0$ mm $\qquad p_y = 275 \text{ N/mm}^2$

*Table 7*                  $\varepsilon = 1.0$    Section Classification
*Figure 3*                 $b = B - 3t = (120 - 3 \times 8.0) = 96$ mm
                           $d = D - 3t = (200 - 3 \times 8.0) = 176$ mm
                           $b/t = 96/8.0 = 12 \quad < 26\varepsilon$        **Flanges are plastic**
                           $d/t = 176/8.0 = 22 \quad < 39\varepsilon$      **Web is plastic**

Section Tables   $A_{gross} = 4830$ mm$^2$
*Clause 4.7.4*      $P_c = A_g p_c$
*Table 24*          $L_e = 0.85 \times 5000 = 4250$ mm

*Clause 4.7.3.2*    $\lambda = \dfrac{L_e}{r_{yy}} = \dfrac{4250}{48.6} = 87.5 \quad \leq 180$

*Table 25*          Use *Table 27(a)*        $p_c \approx 186$ N/mm$^2$
*Table 27(a)*       $P_c = (4830 \times 186)/10^3 = 898$ kN $< 1002$ kN

**Section is inadequate and must be increased in size such that $P_c > 1002$ kN**

Bottom chord and diagonal members have maximum forces of 683 kN and 734 kN in tension respectively.

*Clause 4.6.1*     $P_t = A_e p_y$
Assuming welded connections, then         $A_e = A_{gross}$

$$P_t = (4830 \times 275)/10^3 = 1328 \text{ kN} \quad > 734 \text{ kN}$$
**Section is adequate**

A smaller section can be used, but for aesthetics and ease of detailing connections the breadth '$b$' of all members should be the same.

The reader should redesign the top chord member and design the secondary bracing members CB and BD in a similar manner.


### 3.11  Example 3.7  Concentrically loaded single storey column

A single storey column supports a symmetrical arrangement of beams as shown in Figure 3.29. Using the data provided check the suitability of a 203 × 203 × 60 UC section.

**Design Loads:**
         $F_1 =$  characteristic dead load              $G_k = 75$ kN
         $F_1 =$  characteristic imposed load           $Q_k = 175$ kN
         $F_2 =$  characteristic dead load              $G_k = 20$ kN
         $F_2 =$  characteristic imposed load           $Q_k = 75$ kN

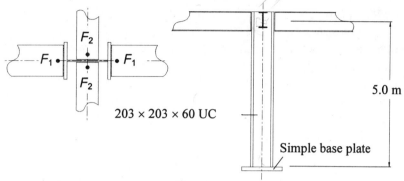

**Figure 3.29**

**Solution:**
Assuming the column to be effectively held in position at both ends, but not restrained in direction at either end, *Table 42* indicates an effective length equal to *L*.

Section Properties:

    $D = 209.6$ mm    $B = 205.8$ mm   $T = 14.2$ mm   $A = 7640$ mm$^2$

    $r_{yy} = 52.0$ mm    $r_{xx} = 89.6$ mm   $t = 9.4$ mm   $d/t = 17.1$  $b/T = 7.25$

| | | |
|---|---|---|
| *Table 6* | $T < 16$ mm   $\therefore p_y = 275$ N/mm$^2$ | |
| *Table 7* | $b/t < 26\varepsilon$ | **Flanges are plastic** |
| | $d/t < 39\varepsilon$ | **Web is plastic** |

*Table 42*      Effective length = 5000 mm

*Clause 4.7.3.2*    $\lambda_{yy} = \dfrac{L}{r_{yy}} = \dfrac{5000}{52.0} = 96 \quad < \quad 180$

                 $\lambda_{xx} = \dfrac{L}{r_{xx}} = \dfrac{5000}{89.6} = 56 \quad < \quad 180$

*Table 25*   For a rolled H-section    x-x axis use *Table 27(b)*
                                      y-y axis use *Table 27(c)*

*Table 27(b)*  $\lambda = 56$  $p_y = 275$ N/mm$^2$        $\therefore p_c = 275$ N/mm$^2$
*Table 27(c)*  $\lambda = 96$  $p_y = 275$ N/mm$^2$        $\therefore p_c = 132$ N/mm$^2$
                                   $\therefore$ critical value $= 132$ N/mm$^2$

*Clause 4.7.4*         $P_c = A_g p_c = (7640 \times 132)/10^3 = 1008$ kN

 Design load   $= 2 \times (F_1 + F_2)$
                 $= 2 \times [(1.4 \times 75) + (1{,}6 \times 175) + (1.4 \times 20) + (1.6 \times 50)]$
                 $= 986$ kN   $< 1008$ kN

                                                **Section is suitable**

### 3.12  Column Baseplates

Columns which are assumed to be nominally pinned at their bases are provided with a slab base comprising a single plate fillet welded to the end of the column and bolted to the foundation with four holding down (H.D.) bolts. The base plate, welds and bolts must be of adequate size, stiffness and strength to transfer the axial compressive force and shear at the support without exceeding the bearing strength of the bedding material and concrete base, as shown in Figure 3.30.

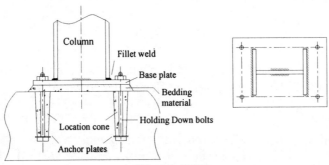

**Figure 3.30**

*Clause 4.13.2.2* of BS 5950:Part 1 gives the following empirical formula for determin-ing the minimum thickness of a rectangular base plate supporting a concentrically loaded column:

$$t = \sqrt{\frac{2.5}{p_{yp}}\, w\left(a^2 - 0.3b^2\right)}$$

where:
| | |
|---|---|
| $a$ | is the greater projection of the plate beyond the column as shown in Figure 3.31 |
| $b$ | is the lesser projection of the plate beyond the column as shown in Figure 3.31 |
| $w$ | is the pressure on the underside of the plate assuming a uniform distribution |
| $p_{yp}$ | is the design strength of the plate from *Clause 3.1.1* or *Table 6*, but not greater than 270 N/mm² |

**Note:** the value of 270 N/mm² is a conversion of the value adopted in the previous permissible stress design code for steelwork BS 449:Part 2: 1969 'The Use of Structural Steel In Building'.

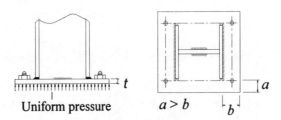

**Figure 3.31**

In addition $t \geq$ the flange thickness '$T$' of the column being supported.

The code indicates that this formula may be used for I, H, channel, box or RHS sections. The use of the equation, which assumes full two-way bending of the plate, can produce plate thicknesses less than would be obtained at critical sections assuming simple cantilever actions. An alternative approach considering a series of simple cantilever projections, which is more appropriate for sections with a large depth/width ratio, e.g. Universal Beams, is the Effective Area Method. This technique is comprehensively explained in Reference 14.

The dimensions of the plate must be sufficient to distribute the axial compressive load to the foundations and to accommodate the holding down bolts.

The purpose of the welds is to transfer the shear force at the base and securely attach the plate to the column. In most cases either 6 mm or 8 mm fillet welds run along the flanges and for a short distance on either side of the web will be adequate.

The holding down bolts are generally cast within location cones in a concrete base and fitted with an anchor plate to prevent pull-out. The purpose of the location cone is to allow for movement before final grouting and hence permit site adjustment during construction. The diameter at the top of the location cones is usually at least 100 mm or 3 × bolt dia-meter. The recommended size of H.D. bolts is M20 for light construction, M24 for bases up to 50 mm thick increasing to M36 for heavier plates. Clearance holes in the base plates should be 6 mm larger than the bolt diameter. The bolts are imbedded in the concrete base to a length equal to approximately 16 to 18 × bolt diameter with a threaded length at least equal to the bolt diameter plus 100 mm.

Bedding material can be either mortar, fine concrete or a proprietary, non-shrink grout. In lightly loaded bases a gap of 25 mm to 50 mm is normally provided; this allows access for grouting the H.D. bolt pockets and ensuring that the gap under the base plate is completely filled. It is normal for the strength of the bedding material to be at least equal to that of the concrete base. In BS 5950: Part 1 *Clause 4.13.1*, the bearing strength for concrete foundations is given as $0.4f_{cu}$, where $f_{cu}$ is the characteristic concrete cube strength at 28 days.

## 3.13 Example 3.7 Slab base

Design a suitable base plate for the axially loaded column in Example 3.6 in which the design axial load :

$$F = 986 \text{ kN} \qquad \text{Assume} \quad f_{cu} = 40 \text{ N/mm}^2$$

$$D = 209.6 \text{ mm} \qquad B = 205.8 \text{ mm} \qquad T = 14.2 \text{ mm}$$

**Solution:**

The minimum size of base plate such that the concrete bearing strength is not exceeded is given by:

$$A_{min} \geq \frac{986 \times 10^3}{0.4 \times 40} = 61.625 \times 10^3 \text{ mm}^2$$

Assuming a square base $\qquad$ Length of base $= \sqrt{61625} = 248 \text{ mm}$

Try a base 300 mm × 300 mm

Actual bearing pressure $\quad w \;=\; \dfrac{F}{\text{Area provided}} \;=\; \dfrac{986 \times 10^3}{300 \times 300} \;=\; 10.96 \text{ N/mm}^2$

greater projection beyond column $\;=\; (300 - 205.8)/2 \;=\; 47.1 \text{ mm}$
lesser projection beyond column $\;=\; (300 - 209.6)/2 \;=\; 45.2 \text{ mm}$

$p_{yp} = 270 \text{ N/mm}^2 \qquad t \;=\; \sqrt{\dfrac{2.5}{p_{yp}} w \left( a^2 - 0.3 b^2 \right)}$

$$=\; \sqrt{\dfrac{2.5 \times 10.96}{270} \left( 47.1^2 - 0.3 \times 45.2^2 \right)} \;=\; 12.8 \text{ mm}$$

Also $\qquad\qquad\qquad t \;\geq\; \text{Flange thickness} = 14.2 \text{ mm} \qquad \text{Assume 15 mm}$

## H.D. Bolts
Assume M20's

threaded length $\;\approx\; 20 + 100 \;=\; 120 \text{ mm}$
imbedded length $\;\approx\; 16 \times 20 \;=\; 320 \text{ mm}$
thickness of bedding material $\;=\; 25 \text{ mm}$
Total length of bolt $\;=\; 120 + 15 + 25 + 320 \;=\; 480 \text{ mm}$

Adopt base plate: 300 mm × 300 mm × 15 mm thick
6 mm fillet welds to flanges and web
M20 mm H.D. bolts 480 mm long with 120 mm threaded length

# 4. Members Subject to Combined Axial and Flexural Loads

## 4.1 Introduction

While many structural members have a single dominant effect, such as axial loading or bending, there are numerous elements which are subjected to both types of loading at the same time. The behaviour of such elements is dependent on the interaction characteristics of the individual components of load. Generally, members resisting combined tension and bending are less complex to design than those resisting combined compression and bending, since the latter are more susceptible to associated buckling effects. The combined effects can occur for several reasons such as eccentric loading or rigid frame action, as illustrated in Figures 4.1(a) and (b) respectively.

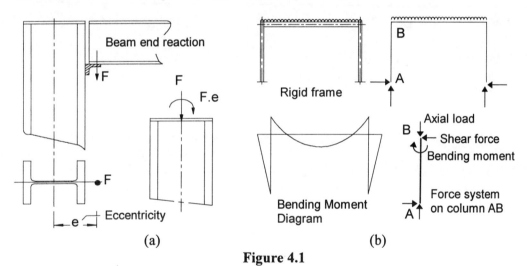

**Figure 4.1**

### 4.1.1 Combined Tension and Bending

Consider a structural member subjected to concentric axial loading as shown in Figure 4.2. The limiting value of applied axial load $F$ can be determined by using the equation:

$$F = A_e p_y \qquad \therefore \quad \frac{F}{A_e p_y} = 1.0$$

Similarly, if the applied load is eccentric to the X-X axis as shown in Figure 4.2(b):

$$M_x = M_{cx} \qquad \therefore \quad \frac{M_x}{M_{cx}} = 1.0$$

In the case of eccentricity about the Y-Y axis, as shown in Figure 4.2(c):

123

$$M_y = M_{cy} \qquad \therefore \quad \frac{M_y}{M_{cy}} = 1.0$$

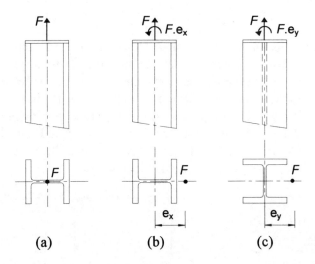

(a)　　　　　　　　(b)　　　　　　　　(c)

**Figure 4.2**

If these limits are plotted on three-dimensional orthogonal axes, then they represent the members' capacity under each form of loading acting singly. Figure 4.3(a) is a linear inter-action diagram. Any point located within the boundaries of the axes and the interaction surface represents a combination of applied loading $F$, $M_x$ and $M_y$ for which

$$\frac{F}{A_e p_y} + \frac{M_x}{M_{cx}} + \frac{M_y}{M_{cy}} \leq 1.0 \tag{1}$$

and which can be safely carried by the section.

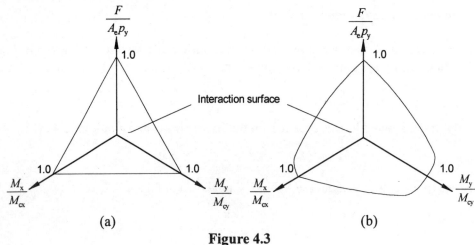

(a)　　　　　　　　　　　　(b)

**Figure 4.3**

Equation (1) which represents a linear approximation of member behaviour is used in BS 5950 in *Clause 4.8.2*. A more rigorous analysis allowing for plastic behaviour of plastic and compact sections results in an interaction surface as shown in Figure 4.3(b). The precise shape of the surface is dependent on the cross-section for which the diagram is constructed. This non-linear surface is represented by:

$$\left(\frac{M_x}{M_{rx}}\right)^{z_1} + \left(\frac{M_y}{M_{ry}}\right)^{z_2} \leq 1.0 \tag{2}$$

where:
$M_{rx}$   the reduced moment capacity about the major axis due to axial loading
$M_{ry}$   the reduced moment capacity about the minor axis due to axial loading
   (see Section 4.1.2 regarding reduced moment capacity)
$z_1$ = 2.0     for I and H, solid and hollow circular sections
   = 1.67   for solid and rectangular hollow sections
   = 1.0     for all other cases
$z_2$ = 2.0     for solid and hollow circular sections
   = 1.67   for solid and rectangular sections
   = 1.0     for all other cases

The values of $M_{rx}$ and $M_{ry}$ are available from published Section Properties Tables. The BS gives equation (2) as a more economic alternative to using equation (1) when designing members subject to both axial tension and bending

**Note:** It is also necessary to check the resistance of such members to lateral torsional buckling in accordance with *Clause 4.3* as shown in Chapter 2, Section 2.4.4, assuming the value of the axial load to be zero.

### 4.1.2  Reduced Moment Capacity

In Chapter 2, Section 2.2 and Figure 2.3 the plastic stress distribution is shown for a cross-section subject to pure bending. When an axial load is applied at the same time, this stress diagram should be amended. Consider an I-section subjected to an eccentric compressive load as shown in Figure 4.4.

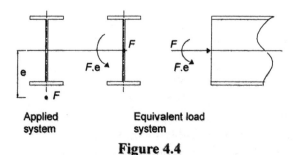

Applied
system

Equivalent load
system

**Figure 4.4**

The equivalent load on the cross-section comprises two components: an axial load '*F*' and a bending moment (*F* × e). The stress diagram can be considered to be the superposition of two components as shown in Figure 4.5.

When the axial load is relatively low then sufficient material is contained within the web to resist the pure axial effects as shown in Figure 4.5(a). There is a small reduction in the material available to resist bending and hence the reduced bending stress diagram is as shown in Figure 4.5(b). The stress diagram relating to combined axial and bending effects is shown in Figure 4.5(c) indicating the displaced plastic neutral axis.

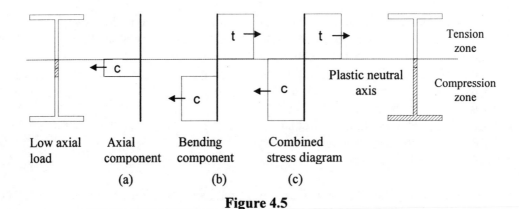

| Low axial load | Axial component | Bending component | Combined stress diagram |
|:--:|:--:|:--:|:--:|
| | (a) | (b) | (c) |

**Figure 4.5**

When the axial load is relatively high then in addition to the web material, some of the flange material is required to resist the axial load as shown in Figure 4.6(a), and hence a larger reduction in the bending moment capacity occurs as shown in Figure 4.6(b). As before the plastic neutral axis is displaced as shown in Figure 4.6 (c).

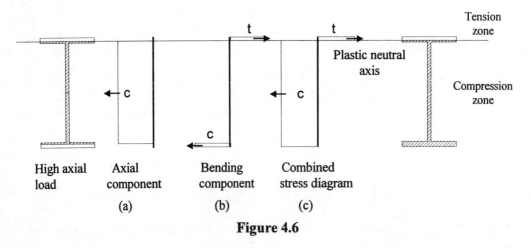

| High axial load | Axial component | Bending component | Combined stress diagram |
|:--:|:--:|:--:|:--:|
| | (a) | (b) | (c) |

**Figure 4.6**

The published Section Property Tables provide a *'change formula'* for each section to enable a reduced plastic section modulus to be evaluated depending on the amount of material required to resist the axial component of the applied loading.

### 4.1.3 Combined Compression and Bending

The behaviour of members subjected to combined compression and bending is much more complex than those with combined tension and bending. A comprehensive explanation of this is beyond the scope of this text and can be found elsewhere (Refs:12, 17). Essentially, there are three possible modes of failure to consider:

(i)     a combination of column buckling and simple uniaxial bending;
(ii)    a combination of column buckling and lateral torsional beam buckling;
(iii)   a combination of column buckling and biaxial beam bending.

As with combined tension and bending, an interaction diagram can be constructed to illustrate the behaviour of sections subjected to an axial load $F$, and bending moments $M_x$ and $M_y$ about the major and minor axes respectively; this is shown in Figure 4.7.

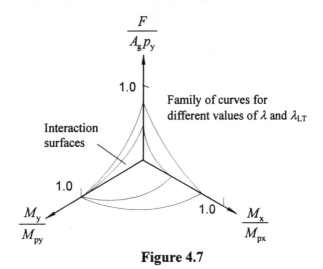

**Figure 4.7**

Although similar, this interaction diagram differs from that shown in Figure 4.3 in that it relates to *slender* members. The precise intercept on the $F$ and $M_x$ axes will depend on the slenderness of the member. Clearly, short, stocky members will intercept the $\dfrac{F}{A_g p_y}$ axis at a value of 1.0; as the slenderness $\lambda$ increases, the intercept decreases. Similarly, in sections which are fully restrained against lateral torsional buckling the intercept on the $\dfrac{M_x}{M_{px}}$ axis will equal 1.0, decreasing as the equivalent slenderness $\lambda_{LT}$ increases.

The minor axis strength is not affected by overall buckling of the member and the intercept is always equal to 1.0.

Members with combined compression and bending should be checked for local capacity using the simplified approach as given in *Clause 4.8.3.2:*

$$\frac{F}{A_g p_y} + \frac{M_x}{M_{cx}} + \frac{M_y}{M_{cy}} \leq 1.0 \tag{3}$$

which differs slightly from the equation given in *Clause 4.8.2* for combined tension and bending, or alternatively using the more exact method:

$$\left(\frac{M_x}{M_{rx}}\right)^{z1} + \left(\frac{M_y}{M_{ry}}\right)^{z2} \leq 1.0 \tag{4}$$

as before.

In addition an overall buckling check should be carried out in accordance with *Clause 4.8.3.3*.

A simplified, conservative approach to the design of members with combined axial compression and biaxial bending is given in *Clause 4.8.3.3.2* when considering overall buckling. The following relationship should be satisfied:

$$\frac{F}{A_g p_c} + \frac{mM_x}{M_b} + \frac{mM_y}{p_y z_y} \leq 1.0 \tag{5}$$

In *Clause 4.8.3.3.2* a more rigorous analysis is used to produce an alternative equation which will generally produce a more economic design. The following relationship must be satisfied:

$$\frac{mM_x}{M_{ax}} + \frac{mM_y}{M_{ay}} \leq 1.0 \tag{6}$$

where:

$M_{ax}$   is the maximum buckling moment about the x-x axis in the presence of axial load and equals the lesser of:

$$M_{cx} \frac{\left(1 - \frac{F}{P_{cx}}\right)}{\left(1 + \frac{0.5F}{P_{cx}}\right)} \qquad \text{or} \qquad M_b \left(1 - \frac{F}{P_{cy}}\right)$$

$M_{ay}$ is the maximum buckling moment about the y-y axis in the presence of axial load and equals:

$$M_{cy} \dfrac{\left(1 - \dfrac{F}{P_{cy}}\right)}{\left(1 + \dfrac{0.5F}{P_{cy}}\right)}$$

where $P_{cx}$ and $P_{cy}$ are the compression resistances about the major and minor axes respectively.

## 4.2  Eccentricity of Loading

The application of loading on members is rarely concentric resulting in pure axial effects. In most cases moments are introduced either due to the eccentricity of loadings caused by eccentric connections, or due to continuity moments in rigid-frame construction.

In the case of angles, channels and T-sections designed as ties, the secondary moment effect induced by the eccentricity of the connection to a gusset plate or other member is allowed for by modifying the effective cross-sectional area, as shown in Chapter 3, Section 3.4 and indicated in *Clauses 4.6.3* and *4.6.4* of BS 5950: Part 1. When designed as struts the effects of secondary moments can be neglected, provided that the members are designed in accordance with *Clause 4.7.10* of the BS.

In general, moments due to eccentricity of connections should be determined and allowed for in the design of members, as discussed in Section 4.1.2; however, in the case of simple construction, e.g. braced frames, *Clause 4.7.6* of BS: 5950 provides guidelines for three separate conditions.

*Condition 1:*
When a beam is supported on a cap plate, as shown in Figure 4.8, the load should be considered to be acting at the face of the column, or edge of packing if used, towards the edge of the beam.

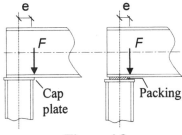

**Figure 4.8**

*Condition 2:*
When a roof truss is supported using simple connections (see Figure 4.9), which cannot develop significant moments, the eccentricity may be neglected and a concentric axial load may be assumed at this point.

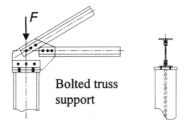

Bolted truss
support

**Figure 4.9**

*Condition 3:*
In all other cases, such as the beam connections shown in Figure 4.10, the eccentricity of loading should be taken as 100 mm from the face of the column or at the centre of the length of stiff bearing, whichever gives the greater eccentricity.
**Note:** The stiff length of bearing of a supporting element is defined in *Clause 4.5.1.3* and *Figure 8.0* of the BS. In most cases the designer does not know the precise details of say, a seating angle, when designing a column and the assumption of 100 mm eccentricity will be adopted.

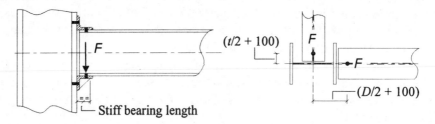

**Figure 4.10**

### 4.3  Section Classification

When a cross-section is subject to combined axial and bending moments the classification of the web should be checked for the web generally (see *Table 7*). For simplicity, the classification may initially be conducted assuming pure axial load; if the result is either plastic or compact, there is no advantage to be gained by carrying out a more complex, accurate calculation. If the section is non-compact, then a more precise calculation should be carried out (refer to Ref:17).

### 4.4  Example 4.1  Industrial unit

An industrial unit comprises a series of braced rectangular frames as shown in Figure 4.11. A travelling crane is supported on a runner beam attached to the underside, at the mid-span point of the rafters. Using the design data provided, check the suitability of a 203 × 133 × 25 UB for the columns of a typical internal frame, using:

(i)     the simplified approach     and
(ii)     the more exact method.

### Design Data:

| | |
|---|---|
| Characteristic dead load due to the sheeting, purlins and services | 0.5 kN/m$^2$ |
| Characteristic imposed load | 0.75 kN/m$^2$ |
| Characteristic dead load due to side sheeting | 0.3 kN/m$^2$ |
| Characteristic dead load due to crane | 5.0 kN |
| Characteristic imposed load | 25.0 kN |

Ignore wind loading

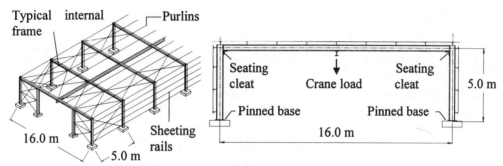

**Figure 4.11**

### Solution:

| | | |
|---|---|---|
| Area of roof supported by typical internal frame = | $(5.0 \times 16.0)$ | = 80 m$^2$ |
| Design dead load due to roof sheeting etc. | = $1.4 \times (0.5 \times 80)$ | = 56 kN |
| Design imposed load on roof | = $1.6 \times (0.75 \times 80)$ | = 96 kN |
| Total distributed design load on rafter | = $(80 + 96)$ | = 176 kN |
| Design load on rafter due to crane | = $(1.4 \times 5.0) + (1.6 \times 25)$ | = 47 kN |
| Area of side sheeting supported by one column = | $(5.0 \times 5.0)$ | = 25 m$^2$ |
| Design dead load due to side sheeting | = $(1.4 \times 25)$ | = 10.5 kN |

Design loads applied to typical internal frame are shown in Figure 4.12

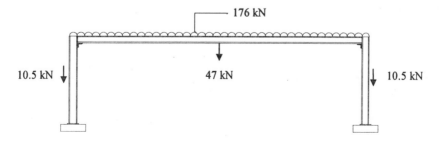

**Figure 4.12**

| Load imposed on column from end of rafter | $= 0.5 \times (176 + 47)$ | $= 111.5$ kN |
|---|---|---|
| Total axial load on column | $= (111.5 + 10.5)$ | $= 122$ kN |

Section Properties: 203 × 133 × 25 UB

| | | | | | | |
|---|---|---|---|---|---|---|
| $D$ = 203.2 mm | $B$ = 133.2 mm | $T$ = 7.8 mm | $Z_{xx}$ = 230 cm$^2$ | $d/t$ = 30.2 |
| $A$ = 3200 mm$^2$ | $d$ = 172.4 mm | $t$ = 5.7 mm | $S_{xx}$ = 258 cm$^2$ | $b/T$ = 8.54 |
| $r_{yy}$ = 31.0 mm | $r_{xx}$ = 85.6 mm | $u$ = 0.877 | $x$ = 25.6 |

*Clause 4 7 6.(a).(3)* An eccentricity equal to ($D/2 + 100$) should be assumed to evaluate the bending moment since the precise details of the seating angles are not known at this stage.

$$\text{Eccentricity} \qquad e = \left( \frac{203.2}{2} + 100 \right) \quad = 201.6 \text{ mm}$$

$$\text{Design bending moment} = (111.5 \times 0.2016) \quad = 22.5 \text{ kNm}$$

| | | |
|---|---|---|
| *Table 6* | $T < 16$ mm | $\therefore \ p_y = 275$ N/mm$^2$ |
| *Table 7* | $8.5\varepsilon \ < \ b/T \ < \ 9.5\varepsilon$ | **Flanges are compact** |

In this case the web should be checked for "web generally" since both bending and axial loads are present, if the section is plastic or compact when considering web to be in compression throughout no precise calculation is required.

*Table 7*

$$\frac{d}{t} < \frac{79\varepsilon}{84} \qquad\qquad \textbf{Web is plastic}$$

*Clause 4.8.3.1*     **Local Capacity Check**

$$\text{Simplified approach} \qquad \frac{F}{A_g p_y} + \frac{M_x}{M_{cx}} + \frac{M_y}{M_{cy}} \quad \leq \ 1.0$$

Since there is no moment about the y-y axis $\dfrac{M_y}{M_{cy}} = 0$

*Clause 4.2.5* $\qquad M_{cx} = p_y S_{xx} = (275 \times 258)/10^3 \quad = 70.95$ kNm

$\qquad\qquad\qquad\quad A_g p_y = (3200 \times 275)/10^3 \qquad\quad = 880$ kN

$$\frac{122}{880} + \frac{22.5}{70.59} \quad = \ 0.46 \ < \ 1.0$$

More Exact Method
$$\left(\frac{M_x}{M_{rx}}\right)^{z1} + \left(\frac{M_y}{M_{ry}}\right)^{z2} \leq 1.0$$

An extract from member capacity tables published by the Steel Construction Institute is given in Table 4.1 in which values are given for $M_{rx}$ and $M_{ry}$. In this case since $M_{ry}$ is equal to zero the second term is not required.

**Table 4.1  Extract from Member Capacity Tables**

| Designation and Axial Load Capacity | F/$P_z$ Semi-Compact (Compact) | Moment Capacity $M_{cx}$, $M_{cy}$ (kNm) and Reduced Moment Capacity $M_{rx}$, $M_{ry}$ (kNm) For Ratio of Axial Load to Axial Load Capacity F/$P_z$ | | |
|---|---|---|---|---|
| | | F/$P_z$ | 0.0 | 0.1 | 0.2 |
| - | - | - | - | - | - |
| | | $M_{cx}$ | 70.9 | 70.9 | 70.9 |
| 203×133×25 | | $M_{cy}$ | 15.2 | 15.2 | 15.2 |
| $P_z = A_g p_y = 879$ | (1.00) | $M_{rx}$ | 70.9 | 69.6 | 65.9 |
| | | $M_{ry}$ | 15.2 | 15.2 | 15.2 |
| - | - | - | - | | - |

Extract from *Steelwork design guide to BS 5950: Part 1: Volume 1* (The Steel Construction Institute)

$$F/P_z = 122/880 = 0.14$$
Using interpolation between the values for 0.1 and 0.2
$$M_{rx} = 68 \text{ kNm}$$

*Clause 4.8.2*
$$z_1 = 2.0$$

$$\left(\frac{M_x}{M_{rx}}\right)^{z1} = \left(\frac{22.5}{68}\right)^{2.0} = 0.11 \ll 1.0$$

*Clause 4.8.3.3.2*  **Overall Buckling Check**

Simplified Approach
$$\frac{F}{A_g p_c} + \frac{mM_x}{M_b} + \frac{mM_y}{p_y z_y} \leq 1.0$$

*Table 18*      $\beta = 0.0 \quad \therefore \quad m = 0.57$

*Table 25*      For I-section use *Table 27(a)* for x-x axis slenderness
use *Table 27(b)* for y-y axis slenderness

*Clause 4.7.3.1*      $\lambda_{xx} = \dfrac{5000}{85.6} = 58.4 \quad < 180$

*Clause 4.7.3.2*                    $\lambda_{yy} = \dfrac{5000}{31.0} = 161.3 \quad < \quad 180$

*Table 27(a)*                       $p_c \leq 241 \text{ N/mm}^2$

*Table 27(b)*                       $p_c \leq 65 \text{ N/mm}^2$

$$p_c = 65 \text{ N/mm}^2$$

*Clause 4.3.7.1*                    $M_b = S_{xx} p_b$

*Clause 4.3.7.7*                    $\lambda = 161.3, \qquad x = 25.6$

*Table 19(b)*                       $p_b \approx 108 \text{ N/mm}^2 \qquad \text{(approximate value)}$

$M_b = (258 \times 108)/10^3 = 27.9 \text{ kNm}$

*Table 18*                          $\beta = 0.0 \qquad\qquad m = 0.57$

As before $M_y = 0.0$

$$\frac{122}{208} + \frac{0.57 \times 10.5}{27.9} = 0.801 \leq 1.0 \qquad\qquad \textbf{Section is adequate}$$

**More Exact Method**              $\dfrac{mM_x}{M_{ax}} + \dfrac{mM_y}{M_{ay}} \quad \leq \quad 1.0$

since $M_y = $ zero                $\dfrac{mM_x}{M_{ax}} \quad \leq \quad 1.0$

$$M_{ax} \leq M_{cx} \frac{\left(1 - \dfrac{F}{P_{cx}}\right)}{\left(1 + \dfrac{0.5F}{P_{cy}}\right)}$$

$M_{cx} = 70.95 \text{ kNm}, \qquad M_b = 27.9 \text{ kNm}$

$P_{cy} = 208 \text{ kN} \qquad\qquad P_{cx} = (3200 \times 241)/10^3 = 771 \text{ kN}$

$$M_{ax} \leq 70.95 \frac{\left(1 - \dfrac{122}{771}\right)}{\left(1 + \dfrac{0.5 \times 122}{771}\right)} = 55.34 \text{ kNm}$$

$$\leq 27.9\left(1 - \frac{122}{208}\right) = 11.54 \text{ kNm}$$

$$M_{ax} = 11.54 \text{ kNm}$$

$$\frac{mM_x}{M_{ax}} = \frac{0.57 \times 10.5}{11.54} = 0.52 \quad < \quad 1.0$$

**Section is adequate**

**Note:** In cases where either $M_x$ or $M_y$ approaches zero the more exact approach may be more conservative (i.e. give a higher value) than the simplified approach. In such cases the values satisfying the simplified approach may be used.

### 4.5 Columns in Simple Multi-Storey Construction (Clause 4.7.7)

A simplified, conservative approach to the design of multi-storey columns in simple construction is given in *Clause 4.7.7* of BS 5950 Part 1. Cost penalties are normally incurred when purchasing and/or transporting sections longer than approximately 18.0 m. Many multi-storey columns are spliced during erection to avoid this or to produce a more economic design by reducing the section sizes used in upper levels. It is normal to consider such columns to be effectively continuous at their splice locations.

In simple construction the beams supported by a column are considered to be simply supported with their end reactions assumed at an eccentricity as indicated in *Clause 4.7.6* of the code and as illustrated in Section 4.2. The net nominal moments induced at any one level by the eccentricities of the beams should be divided between the column lengths above and below that level in proportion to the stiffness, I/L, of each length. When the ratio of the stiffnesses of each length is less than or equal to 1.5 then the net moment may be divided equally. The effects of the nominal moments can be assumed to be limited to the level at which they are applied and do not affect any other level.

When considering loading it may be assumed that all beams supported by a column at any one level are fully loaded. This alleviates the need to consider various load combinations to determine the most critical axial load and moment effects. The following relationship should be satisfied in such columns:

$$\frac{F_c}{A_g p_c} + \frac{M_x}{M_{bs}} + \frac{M_y}{p_y z_y} \leq 1.0$$

where:

$F_c$     the compressive force due to axial load
$p_c$     the compressive strength
$A_g$     the gross cross-sectional area
$M_x$     the nominal moment about the major axis
$M_y$     the nominal moment about the minor axis
$M_{bs}$     the buckling resistance moment for simple columns determined as described for $M_b$ in *Clause 4.3.7.3* and *4.3.7.4* but using an equivalent slenderness ratio $\lambda_{LT}$ of the column given by $\lambda_{LT} = 0.5(L/r_{yy})$.
$p_y$     the design strength

### 4.6 Example 4.2 Multi-storey column in simple construction

The floor plan and longitudinal cross-section of a two-storey, two-bay braced steelwork frame is shown in Figures 4.13 (a) and (b).  Beams B9 and B10 at the roof and first floor level are small tie members, while all others are substantial members that are fixed to the column flanges by web and seating cleats.

   Check the suitability of a 203 × 203 × 52 UC for the column CB2 at section x-x indicated.

**Design Data:**

| | |
|---|---|
| Characteristic dead load at roof level | 4.0 kN/m² |
| Characteristic imposed load at roof level | 1.5 kN/m² |
| Characteristic dead load at first floor level | 6.0 kN/m² |
| Characteristic imposed load at first floor level | 4.0 kN/m² |

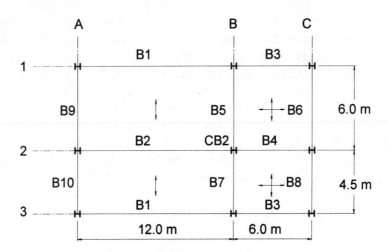

**Figure 4.13(a)**

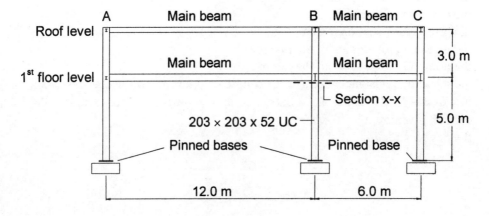

**Figure 4.13(b)**

**Solution:**

Section properties: 203 × 203 × 52 UC

| | | | |
|---|---|---|---|
| $D$ = 206.2 mm | $b/T$ = 8.17 | $I_{xx}$ = 5259 cm$^4$ | $u$ = 0.848 |
| $B$ = 204.3 mm | $d/t$ = 20.4 | $I_{yy}$ = 1778 cm$^4$ | $x$ = 15.8 |
| $t$ = 7.9 mm | $r_{xx}$ = 89.1 mm | $S_{xx}$ = 567 cm$^3$ | $A$ = 6630 mm$^2$ |
| $T$ = 12.5 mm | $r_{yy}$ = 51.8 mm | $Z_{yy}$ = 174 cm$^3$ | $d$ = 160.8 mm |

The column CB2 supports four beams at each level. The slabs between grid-lines A and B are one-way spanning and supported by beams B1 and B2. The slabs between grid-lines B and C are two-way spanning and supported by beams B3 to B8. Beams B2, B4 B5 and B7 impose loads on column CB2 at both the roof and first floor levels. Since the column is the same section throughout, the critical section for design will be section x-x as indicated in Figure 4.13(b). At this location all loading originating from the roof beams is considered to be axial, while the first floor beams will induce nominal moments due to the eccentricity of the connections (see *Clause 4.7.6*) in addition to their axial effects.

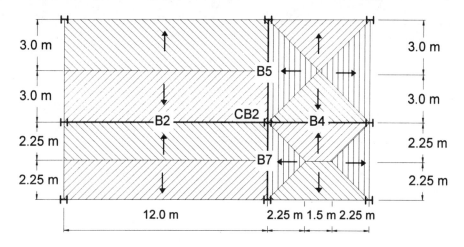

**Figure 4.14**

Area of floor supported by beam B2 $= (6.0 \times 5.25)$ $= 31.5 \text{ m}^2$

Area of floor supported by beam B4 $= \left(\dfrac{6.0}{2.0} \times 3.0\right) + \left(\dfrac{1.5 + 6.0}{2.0}\right)2.25$ $= 17.4 \text{ m}^2$

Area of floor supported by beam B5 $= \left(\dfrac{6.0}{2.0} \times 3.0\right)$ $= 9.0 \text{ m}^2$

Area of floor supported by beam B7 $= \left(\dfrac{4.5}{2.0} \times 2.25\right)$ $= 5.1 \text{ m}^2$

Roof level

Table 2 Design load $= (1.4 \times 4.0) + (1.6 \times 1.5)$ $= 8 \text{ kN/m}^2$

| | | | | | |
|---|---|---|---|---|---|
| Load on beam B2 = | $(31.5 \times 8)$ | = 252 kN | End reaction | = | 126 kN |
| Load on beam B4 = | $(17.4 \times 8)$ | = 139.5 kN | End reaction | = | 69.8 kN |
| Load on beam B5 = | $(9.0 \times 8)$ | = 72 kN | End reaction | = | 36.0 kN |

Load on beam B7 $= (5.1 \times 8) \quad = 40.4$ kN          End reaction $\quad = 20.2$ kN
Axial load transmitted from the roof to column CB2 $\quad = (126 + 69.8 + 36 + 20.2)$
$$= 252 \text{ kN}$$

First Floor level
Table 2          Design load $= (1.4 \times 6.0) + (1.6 \times 4.0) \quad = 14.8$ kN/m$^2$
Load on beam B2 $= (31.5 \times 14.8) \quad = 466.2$ kN          End reaction $\quad = 233.1$ kN
Load on beam B4 $= (17.4 \times 14.8) \quad = 258.1$ kN          End reaction $\quad = 129.1$ kN
Load on beam B5 $= (9.0 \times 14.8) \quad = 133.2$ kN          End reaction $\quad = 66.6$ kN
Design load beam B7 $= (5.1 \times 14.8) = 74.9$ kN          End reaction $\quad = 37.5$ kN

Axial load transmitted from the first floor to column CB2 $=$
$$(233.1 + 129.1 + 66.6 + 37.5)$$
$$= 466.3 \text{ kN}$$
Self-weight of 3.0 m length of column $\quad = (0.52 \times 3.0) \quad = 1.6$ kN
Total axial load at section x-x $\quad = (252 + 466.3 + 1.6) \quad = 720$ kN

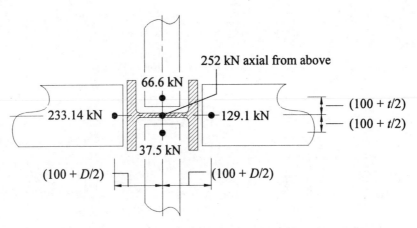

**Figure 4.15**

Net nominal moment about the x-x axis $\quad = (233.1 - 129.1)(100 + \dfrac{206.2}{2})$
$$= 21.13 \text{ kNm}$$

Net nominal moment about the y-y axis $\quad = (66.6 - 37.5)(100 + \dfrac{7.9}{2})$
$$= 3.03 \text{ kNm}$$

*Clause 4.7.7*    Since the column is continuous and the same section above and below
section x-x :

$$\frac{I_{upper}}{L_{upper}} = \frac{I}{3} = 0.3 \qquad \frac{I_{lower}}{L_{lower}} = \frac{I}{5} = 0.2 \qquad \frac{I_u/L_u}{I_L/L_L} = \frac{0.33}{0.2} = 1.65 \ > \ 1.5$$

The net nominal moments applied at the first floor level should be divided in
proportion to the stiffnesses of each length of column.

Lower length at section x-x

$$M_x = 21.13 \times \left(\frac{0.2}{0.2 + 0.33}\right) = 7.97 \text{ kNm}$$

$$M_y = 3.03 \times \left(\frac{0.2}{0.2 + 0.33}\right) = 1.14 \text{ kNm}$$

The load system on the cross-section of the column is equal to the superposition of three components $F$, $M_x$ and $M_y$ as shown in Figure 4.16

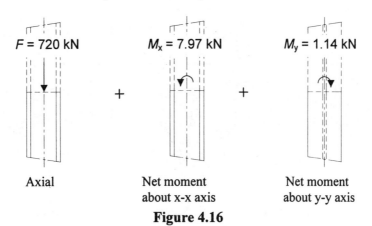

| Axial | Net moment about x-x axis | Net moment about y-y axis |

**Figure 4.16**

*Table 6*      $T = 12.5 < 16$ mm      $\therefore p_y = 275 \text{ N/mm}^2$

Section classification :

*Table 7*      $\varepsilon = 1.0$   $b/T = 8.17 < 8.5\varepsilon$      **Flanges are plastic**
                   $d/t = 20.4$

Since the web is subject to combined axial and bending effects use criterion for the web generally. However, if a check assuming pure axial loading results in a plastic or compact section then precise calculation is not required.

$$d/t < 39\varepsilon \qquad \textbf{Web is plastic}$$

*Clause 4.7.7*      $$\dfrac{F_c}{A_g P_c} + \dfrac{M_x}{M_{bs}} + \dfrac{M_y}{p_y z_y} \leq 1.0$$

$$\lambda_{LT} = 0.5\frac{L}{r_y} = \frac{0.5 \times 5000}{51.8} = 48.3$$

*Table 11*      $\lambda_{LT} = 48.3$,      $p_y = 275 \text{ N/mm}^2$

Interpolating for $\lambda_{LT}$ between 45 and 50

*Design of Structural Steelwork*

Table 4.2 Extract from Table 11 in BS 5950:Part 1

| $\lambda_{LT}$ / $p_y$ | 245 | 265 | 275 |
|---|---|---|---|
| Table 11 Bending strength, $p_b$, (in N/mm$^2$) for rolled sections | | | |
| - | - | - | - |
| - | - | - | - |
| 40 | 238 | 254 | 262 |
| 45 | 227 | 242 | 250 |
| 50 | 217 | 231 | 238 |
| 55 | 206 | 219 | 226 |

$$p_b = 250 - \left(\frac{(250-238)}{5.0} \times (48.3 - 45)\right) = 242.1 \text{ N/mm}^2$$

$$M_{bs} = p_b S_{xx} = (242.1 \times 567)/10^3 = 137.3 \text{ kNm}$$

*Clause 4.7.2.(a)*

Assuming that all the beams provide directional restraint to the column for buckling about both axes, the effective lengths are:

*Table 24*     $L_{ex} = L_{ey} = 0.85 \times 5000 = 4250 \text{ mm}$

$$\lambda_{xx} = \frac{4250}{89.1} = 47.7 \qquad \lambda_{yy} = \frac{4250}{51.8} = 82$$

*Table 25*     rolled **H** sections     x-x axis buckling use *Table 27(b)*
                                        y-y axis buckling use *Table 27(c)*

*Table 27(b)*  $\lambda = 47.4$,   $p_y = 275 \text{ N/mm}^2$,     $\therefore p_c \approx 239 \text{ N/mm}^2$
*Table 27(c)*  $\lambda = 82$,    $p_y = 275 \text{ N/mm}^2$,     $\therefore p_c = 157 \text{ N/mm}^2$

critical value of $p_c = 157$ N/mm$^2$

$$A_g p_c = (6630 \times 157)/10^3 = 1041 \text{ kN}$$
$$p_y z_y = (275 \times 174)/10^3 = 47.85 \text{ kNm}$$

$$\frac{F_c}{A_g p_c} + \frac{M_x}{M_{bs}} + \frac{M_y}{p_y z_y} \leq 1.0 = \frac{720}{1041} + \frac{7.97}{137.3} + \frac{1.14}{47.85} = 0.77 \leq 1.0$$

Section is adequate

## 4.7 Example 4.3 Lattice girder with secondary bending

Consider the lattice girder in Example 3.6, in which a series of castellated beams are supported at the mid-span points between the nodes in the top chord, as shown in Figure 4.17.

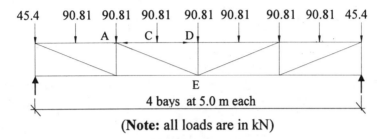

**(Note:** all loads are in kN)

**Figure 4.17**

An alternative to using secondary bracing is to design the top chord members to resist a combined axial load and bending moment. In BS 5950: Part 1, *Clause 4.10* simplified empirical design rules are given for such members in lattice frames and trusses. The axial loads are determined assuming that the connections are pinned (*Clause 4.10(a)*), e.g. the primary axial loads in Example 3.6. The bending moment can be approximated by assuming a 'fixed end moment value' if the load position between the nodes is known, or using *WL*/6 as indicated in *Clause 4.10(c)*.

The effective length of members can be determined taking into account the fixity of the connections and the rigidity of adjacent members.

In this example the axial loads in the lattice girder and the bending moment diagram for the top chord member AD are shown in Figure 4.18.

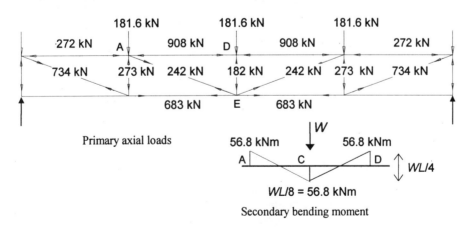

**Figure 4.18**

Member AD :

Design shear force $= (0.5 \times 90.81) = 45.4$ kN

Design axial load                        = 908 kN
Design bending moment                     = 56.8 kNm
Check a 300 × 200 × 10 RHS section for the top chord

Section properties :
   $A$ = 9550 mm$^2$     $d/t$ = 27.0     $b/T$ = 17.0   $r_{xx}$ = 11.2 cm   $r_{yy}$ = 8.14 cm
   $Z_{xx}$ = 796 cm$^3$     $S_{xx}$ = 964 cm$^3$     $t$ = 10 mm

*Table 6*          $t$ = 10 mm   < 16.0 mm        $p_y$ = 275 N/mm$^2$
*Table 7*          $\varepsilon$ = 1.0   Section Classification
*Figure 3*        $b/T$ = 17   < 26$\varepsilon$                              **Flanges are plastic**
                  $d/t$ = 27   < 39$\varepsilon$                              **Web is plastic**

*Clause 4.7.2*     $L_{ex}$ = 0.85 × 5000   = 4250 mm
                  $L_{ey}$ = 1.0 × 2500   = 2500 mm
*Clause 4.7.3*     $\lambda_{xx}$ = $\dfrac{4250}{112}$ = 37.9 < 180,

                  $\lambda_{yy}$ = $\dfrac{2500}{814}$ = 30.7 < 180

*Table 25*     Use *Table 27(a)* for buckling about either the x-x or y-y axis. In this case, the critical slenderness is $\lambda_{xx}$ = 37.9

*Clause 4.8.3.2*  **Local Capacity Check**

Simplified approach          $\dfrac{F}{A_g p_y} + \dfrac{M_x}{M_{cx}} + \dfrac{M_y}{M_{cy}} \leq 1.0$

*Clause 4.2.5*
RHS members have two vertical webs and it is unlikely that the applied shear force will exceed (0.6 × shear capacity) and the moment capacity can normally be taken as :

$$M_c = p_y S_{xx} \leq 1.2 p_y Z_{xx} \text{ for plastic or compact sections}$$

If $S_{xx}$ is greater than $1.2Z_{xx}$ the 1.2 can be replaced by the average load factor.
                  $S_{xx}$ = 964 cm$^3$     $1.2Z_{xx}$ = 1.2 × 796 = 955 cm$^3$
                  $\therefore S_{xx} > 1.2Z_{xx}$

From Example 3.6          Factored load   = 2.36 kN/m$^2$
                          Unfactored load = 1.6 kN/m$^2$

                                   Ratio = $\dfrac{2.36}{1.6}$ = 1.48

$$M_{cx} = p_y S_{xx} = (275 \times 965)/10^3 = 265.4 \text{ kNm}$$
$$\leq 1.48 p_y Z_{xx} = (1.48 \times 796)/10^3 = 324 \text{ kNm}$$
$$M_{cx} = 265.4 \text{ kNm}$$

$$A_g = (9550 \times 275)/10^3 = 2626 \text{ kN}$$
$$M_y = \text{zero}$$

$$\frac{F}{A_g p_y} + \frac{M_x}{M_{cx}} + \frac{M_y}{M_{cy}} \qquad \frac{908}{2626} + \frac{56.8}{265.4} = 0.56 < 1.0$$

Alternatively

*Clause 4.8.3.2(b)*     More Exact Check

$$\left(\frac{M_x}{M_{rx}}\right)^{z1} + \left(\frac{M_y}{M_{ry}}\right)^{z2} \leq 1.0$$

$M_y = $ zero

From published tables with $\dfrac{F}{P_z} = \dfrac{908}{2626} = 0.35$

*Clause 4.8.2*     $z_1 = 5/3$         $M_{rx} = 226 \text{ kNm}$

$$\left(\frac{M_x}{M_{rx}}\right)^{z1} = \left(\frac{56.8}{226}\right)^{\frac{5}{3}} = 0.1 \ll 1.0$$

*Clause 4.8.3.3*     **Overall Buckling Check**
The simplified approach given in *Clause 4.8.3.3.1* assumes that the critical mode of failure is out-of-plane (y-y axis) lateral torsional buckling. In instances where the in-plane buckling (x-x) is more critical, such as in this example, the More Exact Method given in *Clause 4.8.3.3.2* should be used, i.e.

$$\frac{mM_x}{M_{ax}} + \frac{mM_y}{M_{ay}} \leq 1.0$$

where $M_{ax}$ and $M_{ay}$ are as defined previously in section 4.1.3.

*Table 27(a)*     $\lambda_{xx} = 37.9,$     $p_c = 262 \text{ N/mm}^2$
$P_{cx} = A_g p_c = (9550 \times 262)/10^3 = 2502 \text{ kN}$
$\lambda_{yy} = 30.7,$     $p_c \approx 266 \text{ N/mm}^2$
$P_{cy} = A_g p_c = (9550 \times 262)/10^3 = 2540 \text{ kN}$

Buckling resistance moment $M_b$

*Appendix B*        $\lambda_{yy}$  = 30.7,              $D/B$  = $\dfrac{300}{200}$ = 1.5

*Table 38*          1.0  < $D/B$  < 2.0
                    Limiting value of $\lambda$ for   $D/B$  = 1.0  = $\infty$

                    Limiting value of $\lambda$ for   $D/B$  = 2.0  = $\dfrac{350 \times 275}{p_y}$ = 350

Since $\lambda_{yy}$ < 350   Lateral torsional buckling need not be checked and $p_b$ = $p_y$

$$M_b = S_{xx} p_y = 265.4 \text{ kNm}$$
$$M_y = \text{zero}$$

$$M_{cx} \dfrac{\left(1 - \dfrac{F}{P_{cx}}\right)}{\left(1 + \dfrac{0.5F}{P_{cx}}\right)} = 265.4 \dfrac{\left(1 - \dfrac{908}{2502}\right)}{\left(1 + \dfrac{0.5 \times 908}{2502}\right)} = 143.1 \text{ kNm}$$

$$M_b \left(1 - \dfrac{F}{P_{cy}}\right) = 265.4 \left(1 - \dfrac{908}{2540}\right) = 170.5 \text{ kNm}$$

$$M_{ax} = 143.1 \text{ kNm}$$

*Table 18*          $\beta$  = 1.0   $\therefore$ $m$ = 1.0

$$\dfrac{mM_x}{M_{ax}} = \dfrac{1.0 \times 56.8}{143.1} = 0.4 \ < \ 1.0$$

**Section is suitable**

# 5. Connections

## 5.1 Introduction

Traditionally, the consulting engineer has been responsible for the design and detailing of structural frames and individual members, while in many instances the fabricator has been responsible for the design of connections and consideration of local effects. Codes of Practice tend to give detailed specific advice relating to members and relatively little guidance on connection design. This has resulted in a wide variety of acceptable methods of design and details to transfer shear, axial and bending forces from one structural member to another.

Current techniques include the use of black bolts, high-strength friction-grip (H.S.F.G.) bolts, fillet welds, butt welds and, more recently, the use of flow-drill techniques for rolled hollow sections. In addition there are numerous proprietary types of fastener available.

Since fabrication and erection costs are a significant proportion of the overall cost of a steel framework, the specification and detailing of connections is also an important element in the design process.

The basis of the design of connections must reflect the identified load paths throughout a framework, assuming a realistic distribution of internal forces and must have regard to local effects on flanges and webs. If necessary, localised stiffening must be provided to assist load transfer.

All buildings behave as complex three-dimensional systems exhibiting interaction between principal elements such as beams, columns, roof and wall cladding, floors and connections. BS 5950:Part 1, *Clause 2.1.2* specifies four methods of design which may be used in the design of steel frames:

(i) **Simple Design** (*Clause 2.1.2.2*)
In simple design it is assumed that lateral stability of the framework is provided by separate identified elements such as shear walls. portal action, or bracing. The beams are designed assuming them to be pinned at the ends and any moments due to eccentricities of connections are considered as nominal moments when designing the columns.

(ii) **Rigid Design** (*Clause 2.1.2.4*)
In rigid design, full continuity is assumed at connections transferring shear, axial and moment forces between members. In addition it is assumed that adequate stiffness exists at the joints to ensure minimum relative deformation of members and hence maintaining the integrity of the angles between them.

(iii) **Semi-Rigid Design** (*Clause 2.1.2.4*)
In this technique partial continuity is assumed between members. The moment-rotation characteristics of the connection details are used both in the analysis of the framework and the design of the connections. The complexity and lack of readily available data renders this method impractical at the present time.

145

(iv) **Experimental Verification**     *(Clause 2.1.2.5)*

Loading tests may be carried out to determine the suitability of a structure with respect to strength, stability and stiffness if any of the methods (i) to (iii) are deemed inappropriate.

The use of methods (iii) and (iv) will not be considered in this text. The various connection types considered in this chapter are indicated in Figure 5.1.

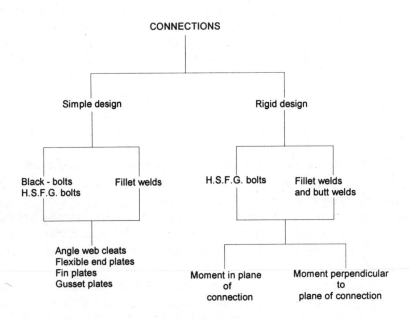

**Figure 5.1**

The design of connections requires analysis to determine the magnitude and nature of the forces which are to be transmitted between members. In both bolted and welded connections this generally requires the evaluation of a resultant shear force and, in the case of moment connections, may include combined tension and shear forces.

### 5.1.1  *Simple Connections*

These are most frequently used in pin-jointed frames and braced structures in which lateral stability is provided by diagonal bracing or other alternative structural elements. Typical examples of instances in the use of simple connections such as in **single or multi-storey braced** frames, or the flange cover plates in beam splices are shown in Figure 5.2.

In each case, the shear force to be transmitted is shared by the number of bolts or area of weld used and details such as end/edge distances and fastener spacings are specified to satisfy the code requirements.

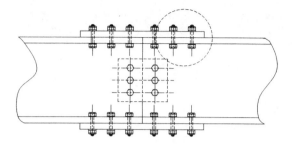

Cover plate to beam flange in beam splice

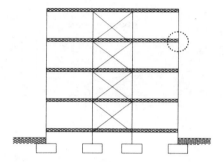

Beam to column connection in braced multi-storey frame

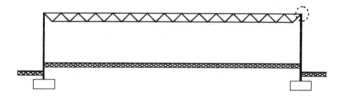

Roof truss to column connection in single-storey braced frame

**Figure 5.2**

### 5.1.2 Moment Connections

These are used in locations where, in addition to shear and axial forces, moment forces must be transferred between members to ensure continuity of the structure. Typical examples of this occur in unbraced single or multi-storey frames, support brackets with the moment either in the plane of or perpendicular to the plane of the connection and web cover plates in beam splices, as shown in Figures 5.3 (a) (b) and (c).

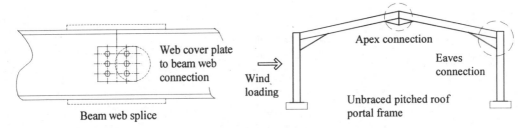

**Figure 5.3(a)**

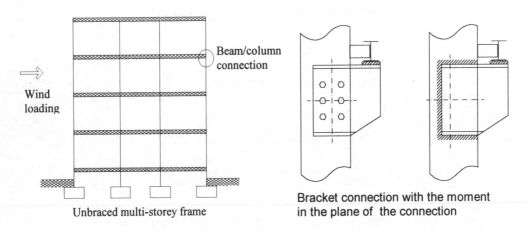

**Figure 5.3 (b)**

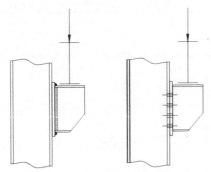

Bracket connections with the moment perpendicular to the plane of the connection

**Figure 5.3 (c)**

A rigorous approach to the design of moment connections is given in the publication *Joints in Steel Construction-Moment Connections* produced by The Steel Construction Institute (Ref: 14).

A simplified more traditional approach is indicated in this text. The following analysis techniques are frequently adopted when evaluating the maximum bolt/weld forces in both types of moment connection.

## 5.1.2.1 *Applied moment in the plane of the connection*

### (i) Bolted connection

The bolts with the maximum force induced in them are those most distant from the centre of rotation of the bolt group and with the greatest resultant shear force when combined with the vertical shear force, i.e. the bolts in the top and bottom right-hand corners.

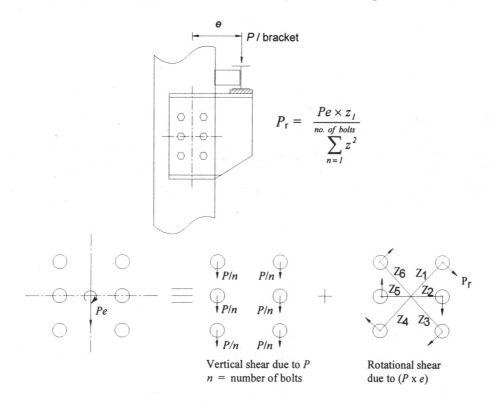

$$P_r = \frac{Pe \times z_1}{\text{no. of bolts} \displaystyle\sum_{n=1} z^2}$$

Vertical shear due to $P$
$n$ = number of bolts

Rotational shear
due to $(P \times e)$

**Figure 5.4**

Total vertical component $\quad F_V = \dfrac{P}{n} + \dfrac{Pe \times z_1}{\displaystyle\sum z^2} \mathrm{Cos}\theta$

Total horizontal component $F_H = \dfrac{Pe \times z_1}{\displaystyle\sum z^2} \mathrm{Sin}\,\theta$

Resultant maximum shear force $= F_R = \sqrt{\left(F_V^2 + F_H^2\right)}$

Similarly for a web splice, e.g. as shown in Figures 5.5(a) and 5.5(b).

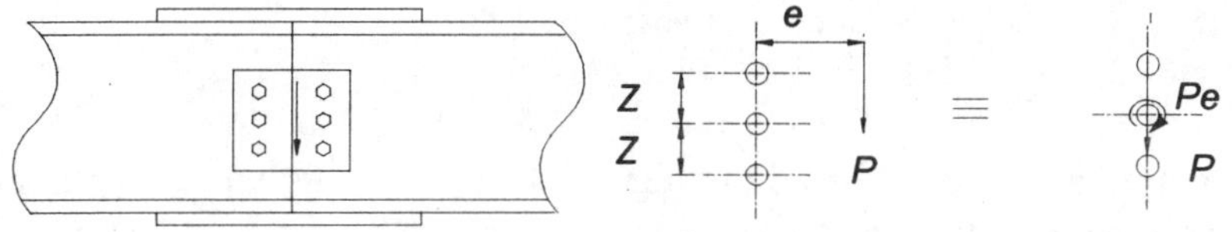

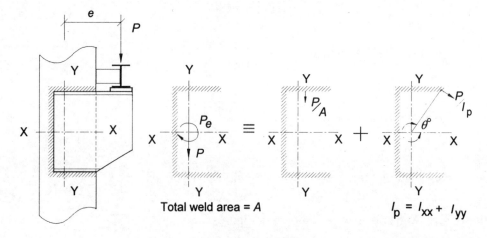

Actual force system                 Equivalent direct shear
                                    and moment

**Figure 5.5(a)**

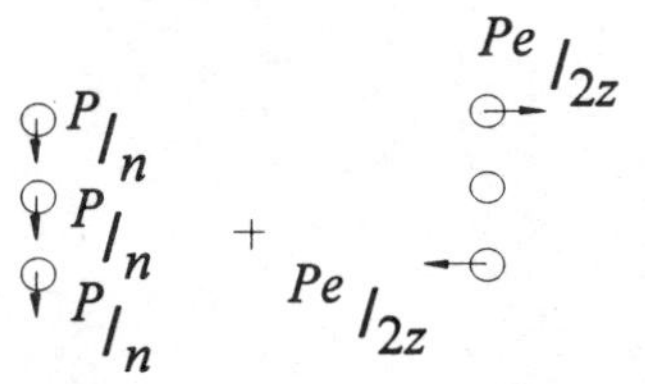

Direct
shear
Rotational
shear

*Individual bolt shear loads*

**Figure 5.5(b)**

$$P_r = \sqrt{\left(\frac{P}{n}\right)^2 + \left(\frac{Pe}{2z}\right)^2}$$

**(ii) Welded connection**

In a welded connection the extreme fibres of the weld most distant from the centre of gravity of the weld group are subjected to the maximum stress values as indicated in Figure 5.6.

Total weld area = A

$I_p = I_{xx} + I_{yy}$

**Figure 5.6**

Total vertical component $\quad F_V \quad = \quad \dfrac{P}{Area\ of\ weld} + \dfrac{Pe}{I_{polar}} Cos\theta$

Total horizontal component $\ F_H \quad = \quad \dfrac{Pe}{I_{polar}} Sin\theta$

Resultant maximum shear force $\quad = \quad F_R = \sqrt{\left(F_V^2 + F_H^2\right)} \quad$ as before.

### 5.1.2.2 *Applied moment perpendicular to the plane of the connection*

#### (i) Bolted connection
In this type of connection it is often assumed that the bracket will rotate about the bottom row of bolts. While this assumption is not necessarily true, it is adequate for design purposes. In this case the bolts will be subjected to combined tension and shear forces as shown in Figure 5.7.

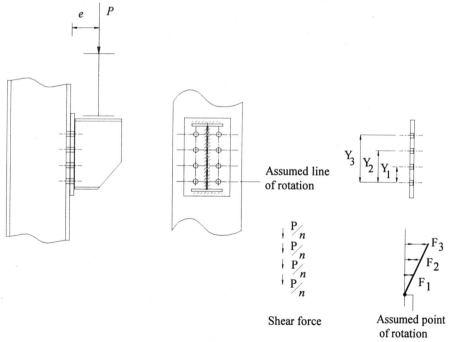

**Figure 5.7**

N = the number of vertical columns of bolts

The Moment of Resistance of the bolt group $\geq$ Applied Moment

$$N \left(F_3 \times y_3 + F_2 \times y_2 + F_1 \times y_1\right) \qquad = P \times e$$

$$N \left(F_3 y_3 + \left[F_3 \frac{y_2}{y_3}\right] y_2 + \left[F_3 \frac{y_1}{y_3}\right] y_1\right) \qquad = Pe$$

$$N \left(\frac{F_3}{y_3}\left[y_3^2 + y_2^2 + y_1^2\right]\right) \qquad = Pe$$

$$F_3 = \frac{Pey_3}{N\displaystyle\sum_{m=1}^{3} y_m^2}$$

where $F_3$ is the maximum tensile load in the bolts which must be designed for the combined the combined effects of the vertical shear and horizontal tension.

(ii)  Welded connection
In welded connections the maximum stress in the weld is determined by vectorial summation of the shear and bending stresses, as shown in Figure 5.8

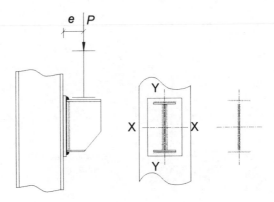

**Figure 5.8**

The maximum vertical shear stress $\qquad F_s = \dfrac{P}{Area\ of\ weld}$

The maximum bending stress $\qquad F_b = \dfrac{P \times e}{Z_{xx}}$

where $Z_{xx}$ is the elastic section modulus of the weld group

Resultant maximum stress $\qquad F_r = \sqrt{F_s^2 + F_b^2}$

Each of these methods is illustrated numerically in the examples in this chapter.

## 5.2 Bolted Connections

The dimensions and strength characteristics of bolts commonly used in the U.K. are specified in BS 3692 (precision hexagon bolts), BS 4190 (black hexagon bolts) and BS 4395 (High-Strength-Friction-Grip bolts). Washer details are specified in BS 4320. BS 4190 specifies two strength grades: *Grade 4.6* which is *mild steel*    (yield stress = 235 N/mm$^2$) and *Grade 8.8*, which is *high-strength-steel* (yield stress = 627 N/mm$^2$).

The most commonly used bolt diameters are: 16, 20, 24 and 30 mm; 22 mm and 27 mm diameter are also available, but are not preferred.

The usual method of forming site connections is to use bolts in clearance holes which are 2 mm larger than the bolt diameter for bolts less than or equal to 22 mm dia., and 3 mm larger for bolts of greater diameter. Such bolts are untensioned and normally referred to as *Black* bolts. In circumstances where slip is not permissible, such as when full continuity is assumed (e.g. rigid-design), vibration, impact or fatigue is likely, or connections are subject to stress reversal (other than that due to wind loading), *High Strength Friction Grip* bolts should be used. Precision hexagon bolts in close tolerance holes have been superseded by H.S.F.G. bolts and are rarely used; they will not be considered further.

### *5.2.1 Black Bolts*

Black bolts transfer shear at the connection by bolt shear at the interface and bearing on the bolts and plates as shown in Figure 5.9

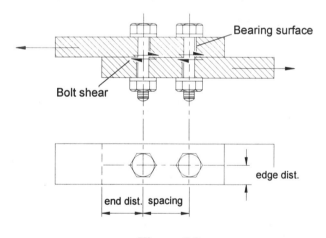

**Figure 5.9**

### *5.2.1.1 Bolt Spacing and Edge Distances*  (*Clause 6.2*)

BS 5950 specifies minimum and maximum distances for the spacing of bolts, in addition to end and edge distances from the centre-line of the holes to the plate edges as indicated. The requirement for minimum spacing is to ensure that local crushing in the wake of a bolt does not affect any adjacent bolts. The maximum spacing requirement is to ensure that the section of plate between bolts does not buckle when it is in compression.

The requirement for minimum end and edge distances is to ensure that no end or edge splitting or tearing occurs and that a smooth flow of stresses is possible. Lifting of the edges between the bolts is prevented by specifying a maximum edge distance.

The values specified in *Clause 6.2* and *Table 31* are illustrated in Figure 5.10.

*5.2.1.2  Bolt Shear Capacity*    (*Clause 6.3.2*)

The shear capacity of a black bolt is given by:

$$P_s = p_s A_s$$

where:

$p_s$    is given in *Table 32* as 160 N/mm$^2$ for Grade 4.6 bolts and 375 N/mm$^2$ for Grade 8.8 bolts

$A_s$    is the cross-section resisting shear, normally this is based on the root of the thread (tensile area $A_t$). If the shear surface coincides with the full bolt shank then the shank area based on the nominal bolt diameter can be used.

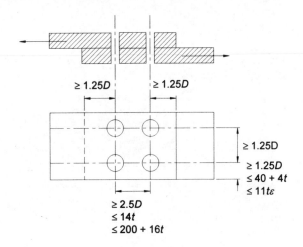

**Figure 5.10**

*5.2.1.3  Bolt Bearing Capacity*  (*Clause 6.3.3.2*)

The bearing capacity of a black bolt is given by

$$P_{bb} = dtp_{bb}$$

where:

$t$    is the thickness of the plate,

$d$    is the nominal bolt diameter, and

$p_{bb}$   is given in *Table 32* as 460 N/mm$^2$ for Grade 4.6 bolts and 1035 N/mm$^2$ for Grade 8.8 bolts.

### 5.2.1.4  Plate Bearing Capacity    (Clause 6.3.3.3)

The bearing capacity of a connected plate is the lesser of:

(i)    $P_{bs} = dtp_{bs}$

or

(ii)    $P_{bs} = 0.5etp_{bs}$

where:
| | |
|---|---|
| $d$ | is the nominal diameter of the bolt, |
| $t$ | is the thickness of the plate, |
| $e$ | is the end distance, |
| $p_{bs}$ | is given in *Table 33* as 460 N/mm² for Grade 43 steel. |

In the case of Grade 4.6 bolts, equation (i) will always be the lesser value provided that the end distance '$e$' is greater than twice the bolt diameter. When Grade 8.8 bolts are used equation (ii) will always be the lesser value.

### 5.2.1.5  Bolt Tension Capacity  (*Clause 6.3.6.1*)

The tension capacity of a black bolt is given by:

$$P_t = p_t A_t$$

where:
| | |
|---|---|
| $A_t$ | is the tensile area of the bolt (based on the root thread diameter) |
| $p_t$ | is given in Table *32* as 195 N/mm² for Grade 4.6 bolts and 450 N/mm² for Grade 8.8 bolts |

### 5.2.1.6  Bolts Subject to Combined Shear and Tension    (*Clause 6.3.6.3*)

When black bolts are subject to both shear and tension simultaneously then the following relationship should be satisfied:

$$\frac{F_s}{P_s} + \frac{F_t}{P_t} \leq 1.4$$

where:

| | |
|---|---|
| $F_s$ | is the applied shear force, |
| $F_t$ | is the applied tensile force, |
| $P_s$ | and $P_t$ are as before |

### 5.2.2  High Strength Friction Grip Bolts  (H.S.F.G.)

H.S.F.G. bolts are manufactured from high strength steel so that they can be tightened to give a high shank tension. The shear force at the connection is considered to be transmitted by friction between the end-plate and column flange plate, as shown in Figures 5.11(a) and (b).

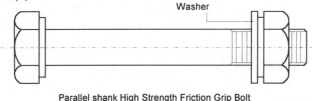

Parallel shank High Strength Friction Grip Bolt

**Figure 5.11(a)**

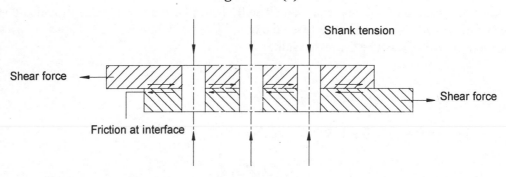

**Figure 5.11(b)**

The bolts must be used with hardened steel washers to prevent damage to the connected parts. The surfaces in contact must be free of mill scale, rust, paint grease etc., since this would reduce the coefficient of friction (**slip factor**) between the surfaces.

It is essential to ensure that bolts are tightened up to the required tension, otherwise slip will occur at service loads and the joint will behave as an ordinary bolted joint. There are several techniques which are used to achieve the correct shank tension; these are now described.

#### 5.2.2.1  Torque Wrench

A power or hand-operated tool which is used to induce a specified torque to the nut.

#### 5.2.2.2  Load-indicating Washers and Bolts

These have projections which squash down as the bolt is tightened. A feeler gauge is used to measure when the gap has reached the required size.

### 5.2.2.3  Part-turning

The nut is tightened up and then forced a further half to three-quarters of a turn, depending on the bolt length and diameter

H.S.F.G. bolts are generally used in clearance holes. The clearances are the same as for ordinary bolts. The design of H.S.F.G. bolts when used in shear, tension and combined shear and tension is set out in *Clause 6.4* of BS 5950:Part 1 and is illustrated below. The possibility of bearing failure must also be considered.

### 5.2.2.4  Shear  (*Clause 6.4.2.1*)

The shear capacity, known as the *slip resistance*, is given by:

$$P_{SL} = 1.1 \, K_S \, \mu \, P_0$$

where:
$P_0$  is the minimum shank tension as specified in BS 4604
  (this is the proof load given in published section tables)
$K_S$  = 1.0 for bolts in clearance holes
$\mu$  = slip-factor, for general grade bolts and untreated surfaces  $\mu$ = 0.45

### 5.2.2.5  Bearing  (*Clause 6.4.2.2*)

The bearing capacity is given by:

$$P_{bg} = dtp_{bg} \leq \frac{etp_{bg}}{3}$$

where:
$d$  = nominal diameter of the bolt,
$t$  = thickness of the connected ply,
$e$  = end distance,
$p_{bg}$  = bearing strength from Table 34.0

### 5.2.2.6  Tension  (*Clause 6.4.4*)

The tension capacity is given by:

$$P_t = 0.9P_0$$

where $P_0$ is as before

### 5.2.2.7  Combined Shear and Tension　　　(Clause 6.4.5)

When H.S.F.G. bolts are subjected to an external tensile force, the clamping action, and hence the friction force available to resist shear, is reduced. Three conditions must be satisfied:

$$\text{(i)} \quad F_S < P_{SL}$$
$$< P_{bg}$$

$$\text{(ii)} \quad F_t < P_t$$

$$\text{(iii)} \quad \frac{F_S}{P_{SL}} + 0.8\,\frac{F_t}{P_t} \leq 1.0$$

## 5.2.3  Design of Simple Connections

The design of simple bolted connections is illustrated in Example 5.1 and Example 5.2.

### 5.2.3.1  Example 5.1  Single lap joint

A lap joint is shown in Figure 5.12 in which a single Grade 4.6  16 mm diameter black bolt is used. There is one shear interface and it is assumed that this passes through the threaded portion of the bolt.

　(a)　Check the minimum and maximum edge and end distances.
　(b)　Determine the shear capacity of the connection with respect to:
　　　(i)　bolt shear,
　　　(ii)　bolt bearing,
　　　(iii)　plate bearing ,　　and
　　　(iv)　plate tension capacity.

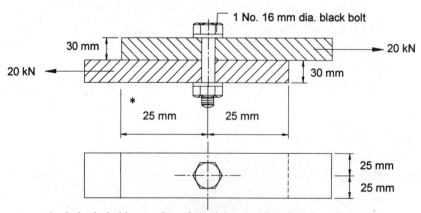

\*　It is desirable to adopt the minimum edge distance +5 mm to accommodate any enlargement which may be necessary on site.

**Figure 5.12**

Solution to Example 5.1 (see Section 5.8)

### 5.2.3.2   *Example 5.2   Double lap joint*

A lap joint similar to Example 5.1 is shown in Figure 5.13; in this case there are two shear interfaces and four 20 mm diameter black bolts. The outer plates are 8 mm thick, while the inner plate is 12 mm thick. Determine (a) and (b) as in the previous example.

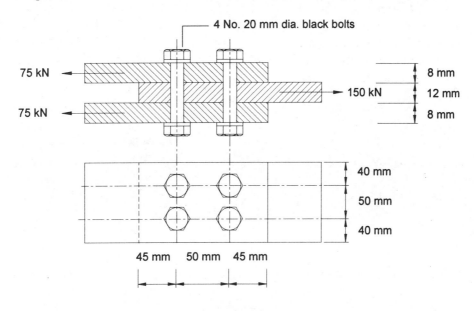

**Figure 5.13**

Solution to Example 5.2 (see Section 5.9)

The use of Grade 8.8 bolts, which have more than twice the shear capacity of Grade 4.6 bolts is more economic, particularly since the installation costs are the same.

## 5.3   Welded Connections

The most common processes of welding used in connections are methods of *fusion* (arc) welding. There are several techniques of fusion welding which are adopted, both manual and automatic/semi-automatic, such as *manual metal arc* (MMA) and *metal inert gas* (MIG). The MMA process is usually used for short runs in workshops or on site and the MIG process for short runs or long runs in the workshop. The two most widely used types of weld are *fillet* and *butt* welds.

### 5.3.1 Fillet Welds

Fillet welds, as illustrated in Figure 5.14, transmit forces by shear through the throat thickness.

The design strength ($p_w$) is given in BS 5950:Part 1, *Table 36*, (in the case of Grade 43 steel $p_w = 215$ N/mm$^2$), and the effective throat thickness as defined in *Clause 6.6.5.3* is normally taken as 0.7 times the effective leg length. It is assumed that when using this value the angle between the fusion faces lies between 60° and 90°. When the fusion faces are inclined at an angle between 91° and 120° then the 0.7 coefficient should be modified as indicated in Figure 5.15:

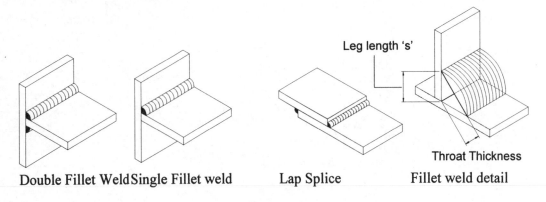

Leg length 's'

Throat Thickness

Double Fillet Weld  Single Fillet weld     Lap Splice          Fillet weld detail

**Figure 5.14**

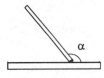

| Angle of fusion faces | Coefficient |
|---|---|
| 91° $\leq \alpha \leq$ 100° | 0.65 |
| 101° $\leq \alpha \leq$ 106° | 0.60 |
| 107° $\leq \alpha \leq$ 113° | 0.55 |
| 114° $\leq \alpha \leq$ 120° | 0.50 |

**Figure 5.15**

In situations when $120° \leq \alpha \leq 60°$ poor access to the acute fillet weld and a small throat thickness on the obtuse fillet weld can create problems. In these situations a different type of weld e.g. a single-sided butt weld, may be more appropriate. BS 5950 indicates in *Clause 6.6.5.4* that the strength of such acute and obtuse fillet welds should be demonstrated by testing.

Where possible a run of fillet weld should be returned around corners for a distance of not less than twice the leg length. If this is not possible, then the length of weld considered effective for strength purposes should be taken as the overall length less one leg length for each length which does not continue round a corner, (*Clause 6.6.5.2*).

In many connections welds are subject to a complex stress condition induced by multi-directional loading. The strengths of transverse fillet welds and longitudinal fillet welds differ, end fillet welds being the stronger. In addition, in side fillet welds, large longitudinal forces are concentrated locally on the member cross-section.

A significant variation in tensile stress occurs across the width of the tensile members when the lateral spacing between weld runs is considerably larger than the length. These effects are limited in BS 5950 by limiting the spacing $T_w$ as indicated in Figure 5.16. In *Clause 6.6.5.5* the code also permits the vectorial summation of stresses to determine the stress for which fillet welds should be designed; this is illustrated in later calculations.

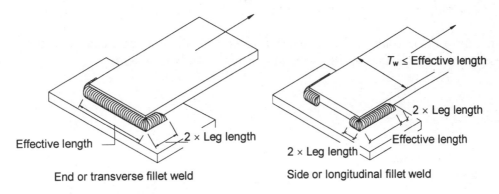

**Figure 5.16**

### 5.3.2 Butt Welds

A butt-weld is formed when the cross-section of a member is fully or partially joined by preparation of one or both faces of the connection to provide a suitable angle for welding. A selection of typical end preparations of butt welds is illustrated in Figure 5.17.

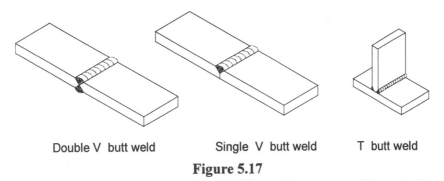

Double V butt weld        Single V butt weld        T butt weld

**Figure 5.17**

The strength of a full penetration butt weld is defined in *Clause 6.6.6.1* and is generally taken as the capacity of the weaker element joined, provided that appropriate electrodes and consumables are used. In most situations when designing structural frames, full penetration welds are unnecessary and the additional costs involved in preparing such connections is rarely justified unless dynamic or fatigue loadings are being considered. Butt welds are used extensively in the fabrication of marine structures such as ships, submarines, offshore oil installations and in pipe-work.

### 5.3.3  Design of Fillet Weld Connections

The design of simple fillet weld connections is illustrated in Example 5.3 and Example 5.4.

#### 5.3.3.1  Example 5.3  Fillet weld lap joint

A flat-plate tie is connected to another structural member as indicated in Figure 5.18. Design a suitable 6 mm side fillet weld to transmit a 75 kN axial force.

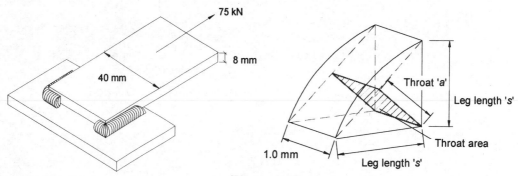

**Figure 5.18**

*Clause 6.6.5*
*Table 36*  Design strength of fillet weld $p_w$ = 215 N/mm²

*Clause 6.6.5.3*

$$\text{Strength of fillet weld /mm length} = p_w \times \text{throat area}$$
$$\text{throat size '}a\text{'} = 0.7 \times \text{leg length '}s\text{'}$$
$$a = 0.7 \times 6.0 = 4.2 \text{ mm}$$
$$\therefore \text{Strength} = (215 \times 4.2 \times 1.0) / 10^3$$
$$= 0.903 \text{ kN}$$

*Clause 6.6.3.2*  Effective length of weld required, $L$ = 75/0.903 = 83 mm

*Clause 6.6.5.2*  Overall length of weld                         = 83+6      = 89 mm
on the side and should be returned round each corner at least equal to 2 × leg length i.e. 12 mm.

$$T_w = 40 \text{ mm}  \quad L > T_w \qquad \textbf{Acceptable}$$

### 5.3.3.2 Example 5.4 Fillet weld - eccentric joint

A $65 \times 50 \times 8$ angle tie is connected by the long leg to a gusset plate and is required to transmit characteristic loads as indicated in Figure 5.19. Design a suitable fillet weld.

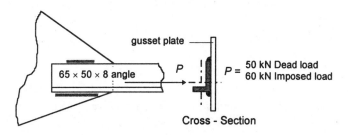

gusset plate

$65 \times 50 \times 8$ angle

$P$

$P = $ 50 kN Dead load
60 kN Imposed load

Cross - Section

**Figure 5.19**

Factored load $= (1.4 \times 50) + (1.6 \times 60) = 166$ kN

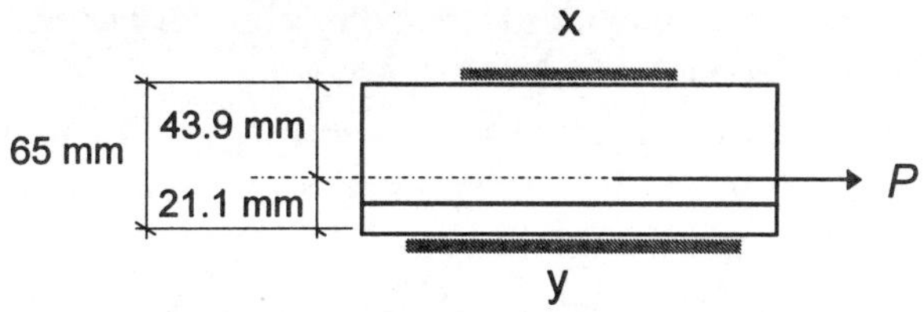

X

65 mm | 43.9 mm | 21.1 mm → $P$

y

**Figure 5.20**

The force is assumed to be transmitted through the centroid of the section Since the force is applied nearer to weld Y than weld X, weld Y will transmit a proportionately larger share of the total force.

*Clause 6.6.5*
*Table 36*              Design strength of fillet weld      $p_w = 215$ N/mm$^2$
*Clause 6.6.5.3*      Assuming a 6 mm fillet weld throat    $a = 0.7 \times 6.0 = 4.2$ mm

Strength of weld /mm length              $= (215 \times 4.2) / 10^3 = 0.9$ kN
Effective length of weld required        $= 166 / 0.9 = 184.4$ mm

$$\text{Weld X} = 184.4 \times \frac{21.1}{65} = 59.9 \text{ mm}$$

$$\text{Weld Y} = 184.4 - 60 = 124.4 \text{ mm}$$

*Clause 6.6.2.1*      End $2 \times$ leg length  i.e.  12 mm
                              Let  X = 75 mm returns $\geq$    and   Y = 140 mm

*Clause 6.6.2.2*                                              **Both X and Y $> Tw$ ($= 65$)**

## 5.4 Beam End Connections

There are three types of beam end connections which are commonly used in the fabrication of steelwork:

- ♦ double angle web cleats,
- ♦ flexible end plates and
- ♦ fin-plates.

The method adopted by any particular fabricator will depend on a number of factors such as the joint geometry and the equipment available. Generally the following characteristics are found:

- ♦ All three types of connection are capable of transmitting at least 75% of the shear capacity of the beam being connected, depending upon the depth of the plates and/or the number of vertical rows of bolts used.
- ♦ Fin plates are the most suitable for the connection of beams which are eccentric to columns or connections which are skewed.
- ♦ End plates are the most suitable when connecting to column webs.
- ♦ Fabrication and treatment does not present any significant problems for any of the three types of connection.

It is important in simple design to detail beam end connections which will permit end rotation of the connecting beam, thus allowing it to displace in a simply supported profile whilst still maintaining the integrity of the shear capacity. This rotational capacity is provided by the slip of the bolts and the deformation of the connection component parts.

### 5.4.1 Double-Angle Web Cleats

Typical angle web cleat comprise 2 / 90 × 90 × 10 angle sections bolted or welded to the web of the beam to be supported as shown in Figure 5.21

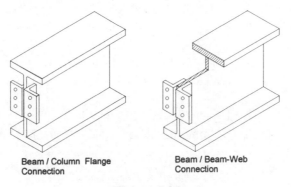

Beam / Column Flange
Connection

Beam / Beam-Web
Connection

**Figure 5.21**

The 10 mm projection of the web cleats beyond the end of the beam is to ensure that when the beam rotates, the bottom flange does not bear on the supporting member. The positioning of the cleats near the top of the beam provides directional restraint to the compression flange. When this positioning is used in addition to a length of cleat equal to approximately 60% of the beam depth, the end of the beam can be assumed to have torsional restraint.

This detail enables an effective length of compression flange of $1.0L$ to be used when designing a beam which is not fully restrained.

### 5.4.2 Flexible End Plates

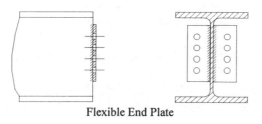

Flexible End Plate

**Figure 5.22**

Connections using flexible end plates are generally fabricated from an 8 mm or 10 mm thick plate which is fillet welded to the web of the beam as shown in Figure 5.22

As with double angle web cleats, the end plates should be welded near the top flange and be of sufficient length to provide torsional restraint. If full depth plates are used which are relatively thick and which may be welded to both flanges in addition to the web, then the basic assumption of a simple connection in which end rotation occurs will be invalid, this could lead to overstressing of the other elements.

### 5.4.3 Fin Plates

A fin-plate connection comprises a length of plate which is welded to the supporting member, such as a column or other beam web, to enable the supported beam to be bolted on site as in Figure 5.23. As with angle cleats and flexible end plates, appropriate detailing should ensure that there is sufficient rotational capability to assume a simple connection.

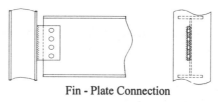

Fin - Plate Connection

**Figure 5.23**

Experimental data has indicated that torsions induced by the plate being connected on only one side of the web of a beam are negligible and may be ignored.

Recommendations for the detailing of fin plates as given in 'Joints in Simple Construction', Vol.2 are :

- ♦ the thickness of the fin plate or beam web should be:
  - (i)  $0.42d$ for grade 50 steel   or
  - (ii)  $0.5d$  for grade 43 steel
- ♦ grade 8.8 bolts are used, un-torqued and in clearance holes,
- ♦ all end and edge distances on the plate and the beam web are at least $2d$,
- ♦ the fillet weld leg length is at least 0.8 times the fin plate thickness,
- ♦ the fin plate is positioned reasonably close to the top flange of the beam,
- ♦ the fin plate depth is at least 0.6 times the beam depth.

### 5.5  Example 5.5  Web cleat, end plate and fin plate connections

A braced rectangular frame in which simple connections are assumed between the columns and rafter beam is shown in Figure 5.24. Using the data provided, design a suitable connection considering;

- (i)      double angle web cleats,
- (ii)     a flexible end plate,
- (iii)    a fin plate

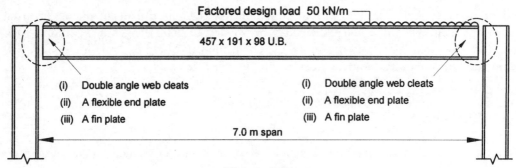

**Figure 5.24**

Assuming simple supports at the ends of the beam:
Total design load on the beam          $= 50 \times 7.0$   $= 350$ kN
Design shear force at the connection                    $= 175$ kN
The use of the following components is assumed where appropriate in each connection;

- (i)      M20 grade 8.8 bolts in 22 mm dia. holes,
- (ii)     6 or 8 mm fillet welds with E43 electrodes,
- (iii)    grade 43 steel adopted throughout.

Solution to  Example 5.5  (see Section 5.10)

## 5.6  Design of Moment Connections

The principle difference between "simple design" and "rigid design" of structural frames occurs in the design of the connections between the elements. In the former the connections are assumed to transmit direct and shear forces, in the latter it is necessary to transmit moments in addition to these forces. The moments can be considered to be either;

(i)     in the plane of the connection as shown in Figure 5.3(a)    or
(ii)    perpendicular to the plane of the connection as shown in Figure 5.3(b)

In both cases the connections are generally designed using either H.S.F.G. bolts or welding. There are a number of approaches to designing connections in moment resisting frames. In most connections the problem is to identify the distribution of forces, moments and stresses in the component parts. Other factors which must be considered are the overall stiffness of the connection,  and the practical aspects of fabrication, erection and inspection.

In rigid frame design each component of the frame can be designed individually to sustain the bending moments, shear forces and axial loads; the connections must then be designed to transfer these forces. In many instances the connections occur where members change direction, such as at the eaves and ridge of pitched roof portal frames. Such frames are usually transported "piece-small" from the fabrication shop to the site and provision must be made for site-joints;  High Strength Friction Grip bolts are ideally suited for this purpose.

In welded connections, such as the knee joint shown in Figure 5.25, because the compressive forces x and y are not collinear, an induced compressive force Z exists to maintain the equilibrium of the forces. This force Z acts across the web plate at the corner and in order to obviate the likelihood of the web plate buckling, corner stiffeners are required as shown, to carry the force Z.

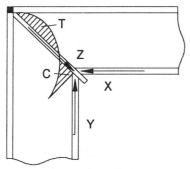

Fillet welds omitted
**Figure 5.25**

The bending stress lines flowing between column and beam are not able to follow a sudden change of direction exactly and become more concentrated towards the inside corner.  This leads to an appreciable shift of the neutral axis towards the inside corner and a redistributed stress diagram, as shown. As the total compression must equal the total tension, an enhanced value of the maximum compressive stress results. This distribution is further modified by the direct compressive stresses in the beam and column.

### 5.8.1  Typical site connection using H.S.F.G bolts

Typical site connections using H.S.F.G. bolts are shown in Figure 5.26. The cap plate transmits the force in the tension flange by means of the H.S.F.G bolts whilst the force in the compression flange is transmitted in direct bearing.  Where the depth of the connection in Figure 26(a) leads to forces of too large a magnitude to be transmitted reasonably, the depth of the connection can be increased by the introduction of a haunch such as is shown in Figure 26(b).

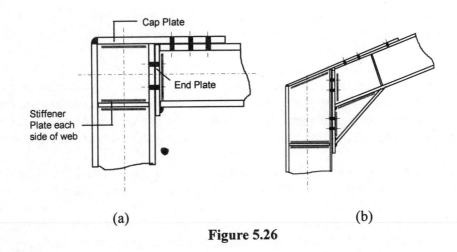

(a)                                        (b)

**Figure 5.26**

Similar details occur at the ridge connection as shown in Figure 5.27

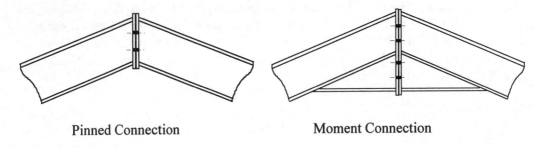

Pinned Connection                    Moment Connection

**Figure 5.27**

### 5.8.2  Example: 5.6        Moment connection in rectangular portal frame

The uniform rectangular portal frame shown in Figure 5.28 is subjected to loading which induces moments and shear forces at the knee joint as given in the data below. Using this data, determine a suitable size of H.S.F.G. bolt for the connection between the column and the roof beam.

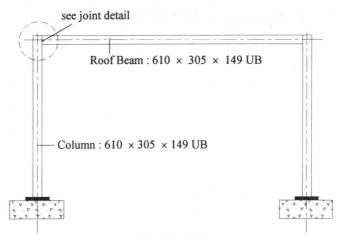

**Figure 5.28**

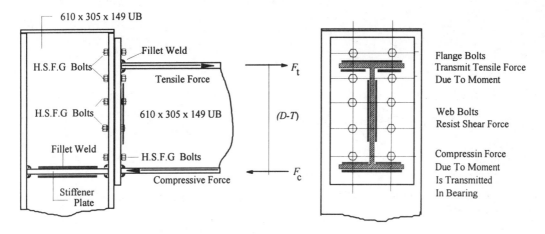

**Figure 5.29**     Column - Beam Joint Detail

Design Data:

| | |
|---|---|
| Column / Beam Section | $610 \times 305 \times 149$ UB |
| Ultimate Design Moment at the knee joint | 540 kNm |
| Ultimate Design Shear Force at the knee joint | 380 kN |

Section Data:     $610 \times 305 \times 149$ UB

| | | |
|---|---|---|
| $D = 609.6$ mm | $d = 537.2$ mm | $I_{xx} = 124.7 \times 10^3$ cm$^4$ |
| $B = 304.8$ mm | $b/T = 7.74$ | $I_{yy} = 9.308 \times 10^3$ cm$^4$ |
| $t = 11.9$ mm | $d/t = 45.1$ | $r_{xx} = 25.6$ cm |
| $T = 19.7$ mm | $r_{yy} = 6.99$ cm | $u = 0.886$ |
| $x = 32.5$ | $A = 190$ cm$^2$ | |

Solution to Example 5.6  (see Section  5.11)

### 5.8.3  Example: 5.7   Crane bracket moment connection

An industrial frame building supports a light electric overhead travelling crane on brackets bolted to the main columns as shown in Figure 5.30 below. Using the design data given determine a suitable size of H.S.F.G. bolt to connect the brackets to the columns.

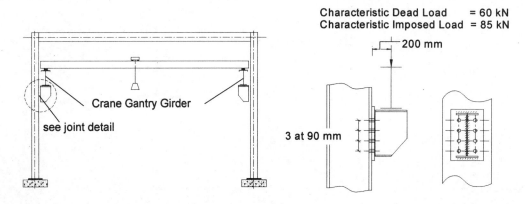

**Figure 5.30**

Assuming the bracket rotates at the level of the bottom bolts (this is conservative) then the

maximum tensile force induced in the bolts   $= T_{max} = \dfrac{P.e.y_{max}}{n\sum y^2}$

where:

n       = number of vertical columns of bolts
e       = eccentricity of applied loads
P       = applied load
$y_{max}$  = distance from the centre of rotation of the bracket to the most distant bolt

$$\sum y^2 = \left(90^2 + 180^2 + 270^2\right) \qquad\qquad = 113.4 \times 10^3 \text{ mm}^2$$

Design Load $= (60 \times 1.4) + (85 \times 1.6)$ $\qquad = 220$ kN
Design Bending Moment $\quad = (220 \times 200)$ $\qquad = 44 \times 10^3$ kNm

$$T_{max} = \frac{44 \times 10^3 \times 270}{2 \times 113.4 \times 10^3} = 52.38 \text{ kN}$$

Maximum Shear Force / bolt $= \dfrac{\text{Design Load}}{\text{Number of Bolts}} = \dfrac{220.0}{8.0} = 27.5$ kN

Assuming 20 mm dia. H.S.F.G. bolts:

$$P_{SL} = 71.3 \text{ kN} \geq 27.5 \text{ kN}$$
$$P_t = 130 \text{ kN} \geq 52.38 \text{ kN}$$

$$\frac{F_S}{P_{SL}} + 0.8\,\frac{F_t}{P_t} = \left[ \frac{27.5}{71.3} + \left(0.8 \times \frac{52.38}{130}\right)\right] = 0.71 \leq 1.0$$

**M20 H.S.F.G.bolts are adequate**

## 5.7 Splices

Fabrication and/or transportation constraints sometimes dictate that beams or columns are delivered on site in separate sections which require spliced connections during the erection stage. In such circumstances it is necessary for the splice to transmit all of the forces, i.e. bending moment, shear and axial forces, which exist at the location of the connection. Spliced connections must provide adequate stiffness and continuity; this is particularly important at locations which are not adjacent to lateral restraints. The tendency for compressive loads to induce lateral instability must be considered and can be accommodated by either:

(i)  providing flange plates of similar dimensions to the flanges of the members being connected,   or

(ii)  satisfying the requirements given in *Appendix C: Clause C.3* of BS 5950:Part: 1 relating to strut action.

The design of splices should satisfy the recommendations given on *Clause 6.1.7* of the code.

### 5.7.1  Beam Splices

Beam splices can be either bolted or welded. Bolted splices using flange and web cover plates as shown in Figure 5.31 (a) and (b) provide more rotational rigidity than welded moment end-plate splices and are the type considered in this text.

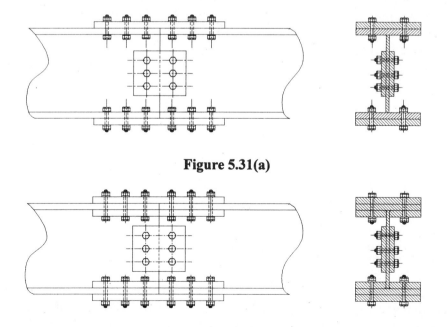

**Figure 5.31(a)**

**Figure 5.30(b)**

Since in most cases, continuity of the beam section has been assumed in the analysis, and joint rotation caused by bolt slip is generally unacceptable, it is recommended that high-strength-friction-grip bolts are used. Usual practice is to adopt a symmetrical arrangement of bolts in each set of cover plates.

Spacing and edge distances as specified in *Clause 6.2* and *6.4* are normally adopted with either M20 or M24 bolts. It is common in the design to assume that the bending moment is resisted by the flange plates, the shear by the web with any co-existent axial load being divided equally between the flanges; this is illustrated in Figures 5.32 (a), (b) and (c).

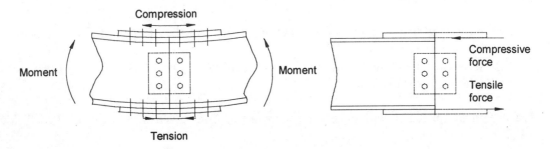

**Figure 5.32(a)**

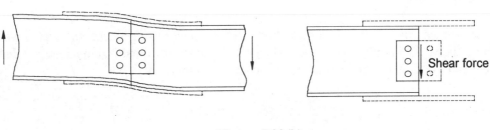

**Figure 5.32(b)**

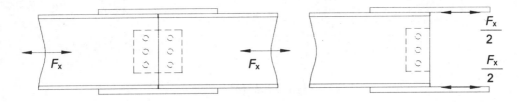

**Figure 5.32(c)**

### 5.7.2  *Example 5.8  Beam splice*

A beam in a rigid-jointed frame structure is required to be site bolted to a welded stub on a column flange as shown in Figure 5.33. Design a suitable spliced connection. Assume that the bending moment induces compression in the top flange of the beam at this location and a compressive axial force is also present.

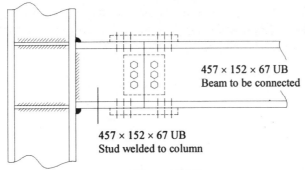

457 × 152 × 67 UB
Beam to be connected

457 × 152 × 67 UB
Stud welded to column

**Figure 5.33**

Design shear force      =  250 kN
Design bending moment =  100 kN
Design axial load      =  75 kN

Section properties: 457 × 152 × 67 UB
    $D$ =  458 mm      $T$ =  15.0 mm     $r_{yy}$ =  32.7 mm    $A$ =  85.6 cm$^2$
    $B$ =  153.8 mm    $t$ =  9.0 mm      $S_{yy}$ =  187 cm$^3$

Force in top flange:

Axial load $= \dfrac{M_x}{(D-T)} + \dfrac{F_x}{2}$ $= \left(\dfrac{100 \times 10^3}{458 - 15} + \dfrac{75}{2}\right)$ $=$ 263.2 kN  compression

Force in bottom flange:

Axial load $= \dfrac{M_x}{(D-T)} - \dfrac{F_x}{2}$ $= \left(\dfrac{100 \times 10^3}{458 - 15} - \dfrac{75}{2}\right)$ $=$ 188.2 kN  tension

**Check beam flanges:**
The beam flange plates must be of sufficient size to transmit the greater of the above two forces allowing for a reduced cross-sectional area due to the bolt holes.

Strength of flange plate   = $p_y A_e \geq$ 263.2 kN
*Clause 3.3.3*          $A_e = K_e \times A_{net} \leq$  gross area
Assuming Grade 43 steel and M20 H.S.F.G. bolts
                $K_e$ =  1.2  and  $A_{net}$  =  [153.8 − (2 × 22)] × 15   =  1647 mm$^2$
*Table 6*            $T$ <  16.0 mm  $p_y$  =  275 N/mm$^2$
Strength of flange plates =  (275 × 1.2 × 1647)/10$^3$  =   543.5 kN     >  263.2 kN
                                            **Flange plates are adequate**

**Design of flange cover plates:**

Minimum area required for cover plates      $\geq \dfrac{263.2 \times 10^3}{275} =$  957 mm$^2$

$$A_e \quad = \quad 1.2 A_{net} \quad = \quad 957 \quad \therefore \quad A_{net} \quad = \quad 796 \text{ mm}^2$$

Assuming cover plates 150 mm wide

$$A_{net} \quad = \quad [150 - (2 \times 22)] \times T_{cover\ plate} \quad = \quad 796$$
$$T_{cover\ plate} \quad \geq \quad 7.5 \text{ mm.}$$

**Adopt 150 mm × 15 mm thick cover plates**

Since the cover plates are the same as the beam flanges, the possibility of failure due to strut action need not be checked; however the calculations have been carried out to illustrate the method.

The equation given in *Appendix C: Clause C.3 :* can be used to evaluate the secondary moment

$$M_{max} = \frac{\eta f_c s}{\left(1 - \dfrac{f_c}{p_c}\right)}$$

where:

$M_{max}$ is the moment about the y-y axis to be shared by both cover plates

$\eta \quad = 0.001 a(\lambda - \lambda_o) \quad \geq 0$

$a \quad = 3.5$

$\lambda \quad$ slenderness $L_e/r_{yy}$

$s \quad$ is the plastic modulus

$$\lambda_o \quad = \quad 0.2\left(\frac{\pi^2 E}{p_y}\right)^{\frac{1}{2}} \quad = \quad 0.2\left(\frac{\pi^2 \times 205 \times 10^3}{275}\right)^{\frac{1}{2}} \quad = \quad 17.15$$

$$\eta \quad = \quad 0.001 \times 3.5 \times (122 - 17.15) \qquad = 0.367$$

$$f_c \quad = \quad \frac{263.2 \times 10^3}{85.6 \times 10^2} \qquad\qquad = 30.75 \text{ N/mm}^2$$

$$S_{yy} \quad = \quad 187 \times 10^3 \text{ mm}^3$$

$$p_e \quad = \quad \frac{\pi^2 \times 205 \times 10^3}{122^2} \qquad\qquad = 135.94 \text{ N/mm}^2$$

$$M_{max} \quad = \quad \frac{0.367 \times 30.75 \times 187 \times 10^3}{\left(1 - \dfrac{30.75}{135.94}\right) \times 10^6} \qquad = 2.73 \text{ kNm}$$

$$= 1.37 \text{ kNm per flange}$$

The flange cover plates should be checked for the combined axial load due to the primary moment and applied axial load, and the secondary moment equal to $M_{max} / 2$ due to strut action.

*Clause 4.8.3.3.1* $\qquad\qquad \dfrac{F}{A_e p_y} + \dfrac{m M_y}{Z_{net} p_y}$

where:

$F$      the applied axial load,

$m$     the equivalent uniform moment factor (assumed to equal 1.0 in this case),

$M$    the applied moment about the y-y axis due to strut action,

$A_e$    the net cross-sectional area of the cover plate,

$Z_{net}$   the elastic section modulus of the cover plate allowing for the holes,

$p_y$    the yield stress from *Table 6*.

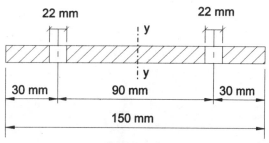

**Figure 5.33**

$$I_{yy} = \frac{15 \times 150^3}{12} - 2\left(\frac{15 \times 22^3}{12} + \left[15 \times 22 \times 45^2\right]\right) = 2.86 \times 10^6 \text{ mm}^4$$

$$Z_{net} = \frac{3.56 \times 10^6}{87.5} = 38.13 \times 10^3 \text{ mm}^3$$

$$A_{net} = (150 \times 15) - (2 \times 22 \times 15) = 1590 \text{ mm}^2$$

$$\frac{F}{A_e p_y} + \frac{mM}{Z_{net} p_y} = \frac{263.2 \times 10^3}{(1590 \times 275)} + \frac{1.37 \times 10^6}{(38.13 \times 10^3 \times 275)} = 0.6 + 0.13 = 0.73$$

**Flange cover plates are adequate**

### Design of flange bolts:

The force in the flange plate bolts is induced by the force due to the bending moment and the co-existent axial load combined with the secondary effects of strut action, i.e.

Shear due to bending moment and axial load      = 263.2 kN

Shear due to strut action      = $(1.37 \times 10^3)/90$   = 15.2 kN

Total shear to be transmitted      = $(263.2 + 15.2)$   = 278.4 kN

Assuming M20 H.S.F.G. bolts,

$$\text{Slip resistance} = P_{SL} = 1.1 \, K_s \mu P_o$$

*Clause 6.4.2.1*        $\text{Bearing resistance} = P_{bg} = d t p_{bg}$

Assume that the end distance      $e \geq 3d$, with 10 mm thick cover plates

$$P_{bg} = d t p_{bg} = (20 \times 10 \times 825)/10^3 = 165 \text{ kN/bolt}$$

The number of bolts is governed by the slip resistance

Minimum number of bolts required  $\geq \dfrac{278.4}{71.3} = 3.9$

**Adopt eight M20 H.S.F.G. bolts with two pairs on each side**

It is normal practice to make both flanges the same.

*Clause 6.2.1*          Minimum spacing  $\geq 2.5d$   $= 2.5 \times 20$  $= 50$ mm
                         Maximum spacing  $\leq 14t$   $= 14 \times 10$   $= 140$ mm
*Clause 6.2.3*   Minimum end/edge distances $\geq 1.25D = 1.25 \times 22 = 27.5$ mm
*Clause 6.2.4*        Maximum edge distance $\leq 11t\varepsilon$   $= 11 \times 10$   $= 110$ mm

In H.S.F.G. bolted connections, the outer splice (i.e. cover plates) should not be thinner than the lesser of:

(i) bolt diameter/2,  or
(ii) 10 mm

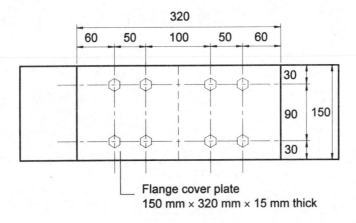

Flange cover plate
150 mm × 320 mm × 15 mm thick

**Figure 5.35**

*Clause 6.4.2.3:* When the distance, $L_j$, between the first and last rows of fasteners (200 mm in the case above), is greater than 500 mm, then the slip resistance of the bolts should be determined using:

$$P_{SL} = 0.6 \times P_o \left( \frac{5500 - L_j}{5000} \right)$$

$$\leq 1.1 K_s \mu P_o$$

$$\leq dt p_{bg} \leq \frac{1}{3} e t p_{bg}$$

**Design of web cover plate:**
The web cover plates should be designed for shear and bending induced by the vertical shear force at the splice. The bolts, which will be in double shear, should be designed to resist the vertical shear combined with the rotational shear caused by the eccentricity of the force, as shown in Figure 5.36.

Cover plate size

| | | |
|---|---|---|
| Minimum edge/end distances | $\geq 1.25D$ | $= 25$ mm |
| end distance to ensure that slip resistance governs | $\geq 3D$ | $= 60$ mm |
| For H.S.F.G. bolts the minimum plate thickness | $\geq D/2$ | |
| | $\geq 10$ mm | $= 10$ mm |

Try a plate 240 mm wide × 300 mm deep × 10 mm thick.

$$\text{Applied shear force} \quad F_v = 250 \text{ kN} = 125 \text{ kN/plate}$$

*Clause 4.2.3*
$$P_v = 0.6 p_y A_v$$
where:
$$A_v = 0.9[(10 \times 300) - (3 \times 22)] = 2640 \text{ mm}^2$$
$$P_v = (0.6 \times 275 \times 2640)/10^3 = 436 \text{ kN} > 125 \text{ kN}$$
**Cover plates are adequate in shear**

A conservative estimate of the bending strength of the web cover plate can be made assuming that:
$$M_b = S_{xx} p_b$$

where $\lambda_{LT}$ used to determine $p_b$ is calculated using the equation given in *Appendix B: Clause B.2.7*,

$$\lambda_{LT} = n \times 2.8 \times \left( \frac{L_e}{t^2} \right)^{\frac{1}{2}}$$

where:
$$n = 1.0 \quad L_e \approx 300 \text{ mm} \quad d \approx 120 \text{ mm} \quad t = 10 \text{ mm}$$

$$\lambda_{LT} = 1..0 \times 2.8 \times \left( \frac{300 \times 120}{10^2} \right)^{\frac{1}{2}} = 53$$

*Table 19*
$$p_b = 231 \text{ N/mm}^2$$

$$S_{xx} = \frac{10 \times 300^2}{4} - 2 \left[ (10 \times 22 \times 90) + (10 \times 22 \times 5.5) \right] = 183 \times 10^3 \text{ mm}^3$$

$$M_b \;=\; (231 \times 183 \times 10^3)\,/\,10^6 \;=\; 42.3 \text{ kNm.}$$

Applied bending moment $\;=\; (250 \times 60)\,/\,10^3 =\; 15$ kNm $=\; 7.5$ kNm / plate

$$M_x \;<\; M_b$$

**Cover plate is adequate in bending**

**Bolts in web cover plate:**

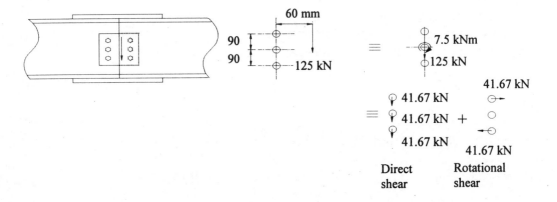

**Figure 5.36**

Resultant shear force $= \sqrt{41.67^2 + 41.67^2} \;=\; 58.9$ kN

$<$ Double shear capacity of the bolts

**Bolts in web are adequate**

### *Column splices*

As with beam splices, continuity of a structure must be maintained at the location of the splice. It is necessary to ensure adequate stiffness about both axes in addition to ensuring adequate strength. In braced multi-storey frames, the nominal moments induced by incoming beams (see Section 4.2, Chapter 4) as indicated in *Clause 4.7.6* must be included in addition to the axial load when designing the splice. The bending moments are assumed to be carried solely by the flanges while the axial load may be shared between the web and the flanges in proportion to their cross-sectional areas.

Although not essential, it is usual to assume that the column ends are prepared for contact in bearing and that untorqued bolts in clearance holes are used.

Generally, column splices are situated a short distance, e.g. 500 mm above floor level. In circumstances in which a splice is located away from a point of lateral restraint, the effects of strut action as illustrated in Section 5.6.1 for beam splices must be considered.

It is advisable wherever possible to minimize secondary effects due to eccentricity and splices should therefore connect members in line such that the centroidal axes of the splice plates coincide with those of the column section above and below the splice. Columns of the same serial size but of different mass per metre run and of different serial sizes can be accommodated by the use of packing plates, web plates and angle sections, as shown in Figure 5.37.

In circumstances in which it is necessary for the splice to comply with structural integrity requirements as specified in *Clause 2.4.5.3*, the flange cover plates and bolts should be capable of transmitting a tensile force equal to two-thirds of the factored vertical load applied to the column floor immediately below the splice. If there is no net tension due to the combined moment and axial load, then the splice can be designed to transmit the applied compressive load by bearing.

The empirical detailing requirements indicated should ensure that sufficient stiffness and robustness of a splice is achieved when only bearing is being considered. If net tension does exist when considering combined moment and axial loading then the strength of the flange cover plates must be checked in tension and bearing, and the bolt group checked in shear.

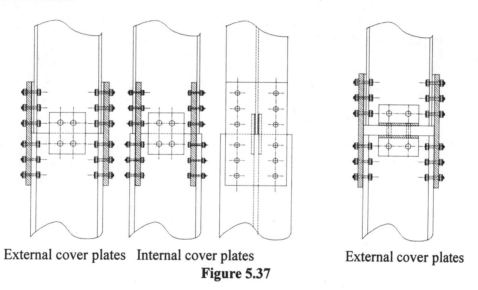

External cover plates    Internal cover plates                    External cover plates
**Figure 5.37**

The following are recommended splice details to ensure adequate stiffness, robustness and continuity.

- The thickness of the flange cover plates should be at least equal to the greater of 10 mm or half the upper column flange thickness.
- When cover plates are positioned on the outside, their width should be at least equal to the width of the upper column flanges.
- The length of the cover plates should be at least equal to the greater of
    (i)  225 mm + division plate thickness        or
    (ii) the width of the upper column flanges
- When columns of the same serial size are used a nominal web cover plate of width at least equal to half the overall depth of the upper column should be used.
- When columns of different serial sizes are used, a division plate of sufficient thickness to allow a 45° dispersion of load from the upper column to the lower column should be used in conjunction with a pair of nominal web cleats.

♦ Grade 8.8, untorqued bolts are normally adopted with typically 4 No. M20 bolts in the web plates.

♦ When combined bending and compressive loading results in a net tensile force being induced, then torqued H.S.F.G. bolts should be used with the cover plates being checked for bearing and tension.

♦ *Clause 6.2* governs the fastener spacing and edge distances.

♦ The spacing of the flange cover plate bolts perpendicular to the direction of stress should be maximized to optimize the joint rigidity.

♦ Packing pieces of steel will be required to accommodate different thicknesses of flange and web plates

### 5.7.4  Example 5.9  Column splice

A column in a multi-storey frame is spliced at a floor level as shown in Figure 5.38. Design a suitable splice which must satisfy structural integrity requirements in addition to transferring the given ultimate loading.

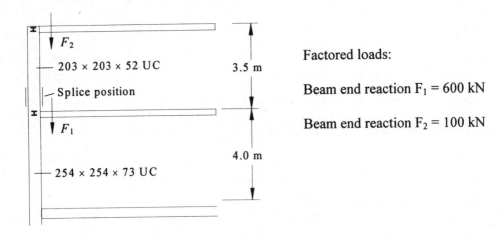

**Figure 5.38**

Section properties: $203 \times 203 \times 52$ UC
$D = 206.2$ mm; $B = 203.9$ mm $T = 12.5$ mm $I_{xx} = 5259$ cm$^4$

Section properties: $254 \times 254 \times 73$ UC
$D = 245.1$ mm; $B = 254.1$ mm $T = 14.2$ mm $I_{xx} = 11410$ cm$^4$

Neglect the self-weight of the tie beams
Beam end reactions due to ultimate loading $\qquad F_1 = 600$ kN
$\qquad\qquad\qquad\qquad\qquad\qquad\qquad\qquad\qquad F_2 = 100$ kN

Design axial loading below the level of the splice $\quad = (100 + 600) = 700$ kN

Design axial loading above the level of the splice $\quad = 100$ kN

*Clause 4.7.6* $\qquad$ Bending moment due to floor beam reaction $F_1 = (F_1 \times$ eccentricity)
$$= (600 \times (0.127 + 0.1)) = 136.2 \text{ kNm}$$

*Clause 4.7.4* $\quad \left(\dfrac{I}{L}\right)_{upper} = \dfrac{5259}{350} = 15 \qquad \left(\dfrac{I}{L}\right)_{lower} = \dfrac{11410}{400} = 28.5$

$$\dfrac{\left(\dfrac{I}{L}\right)_{lower}}{\left(\dfrac{I}{L}\right)_{upper}} = \dfrac{28.5}{15} = 1.9 \geq 1.5$$

The moment should be divided in proportion to the stiffness of the columns.

Design bending moment on lower column $= 136.2 \times \dfrac{28.5}{(28.5 + 15)} = 89.21$ kNm

Design bending moment on upper column $= 136.2 \times \dfrac{15}{(28.5 + 15)} = 47.0$ kNm

*Clause 4.5.1.3* $\qquad$ Assume a 45° dispersion of the load through the division plate
Thickness of plate required $\approx \dfrac{D_1 - D_2}{2} = \dfrac{254 - 206}{2} = 24$ mm

$\qquad\qquad\qquad\qquad\qquad\qquad\qquad\qquad$ **Adopt division plate 25 mm thick**

Check for the presence of tension in the splice due to combined axial load and bending moment

Axial load induced in cover plates by the lower moment $\approx \dfrac{89.21 \times 10^3}{245.1} = 364$ kN

Axial load induced in cover plates by the upper moment $\approx \dfrac{15 \times 10^3}{206.2} = 228$ kN

Since this value is greater than the applied compressive force of 100 kN, tension is developed and it is necessary to check the tensile and bearing capacities of the flange cover plates and the shear capacity of the bolt group. In addition, *Clause 6.3.4* relating to long joints must be satisfied.

**Flange cover plates:**
*Clause 4.6.1:* Tensile capacity $P_t = A_e p_y$
*Table 6:* Since thickness of plates $\leq 16$ mm $\quad p_y = 275$ N/mm$^2$

*Clause 3.3.3:* $\quad A_e = K_e \times$ net area $\quad$ where $\quad K_e = 1.2$
$\qquad\qquad\qquad\qquad \leq$ gross area

| | | |
|---|---|---|
| gross-area | $= 205 \times 10$ | $= 2050$ mm$^2$ |
| net-area | $= 2050 - (2 \times 22 \times 10)$ | $= 1610$ mm$^2$ |
| $K_e \times$ net-area | $= 1.2 \times 1610$ | $= 1932$ mm$^2$ |
| | $P_t = (1932 \times 275)/10^3$ | $= 531$ kN |

Tensile force in flange plate $\approx (228 - 100) = 128$ kN $< 531$ kN
**Cover plates are adequate in tension**

*Clause 6.3.3.3:* $\quad$ Bearing capacity $= dtpb_s \leq 0.5etpb_s$
*Table 33* $\qquad p_{bs} = 460$ N/mm$^2$ $\quad$ Assume that $e \geq 2d$
$\qquad\qquad$ Bearing capacity $= (20\times 10 \times 460)/10^3 = 92$ kN/bolt
$\qquad\qquad$ Bearing capacity $= 6 \times 92 \qquad = 552$ kN $> 128$ kN
**Cover plates are adequate in bearing**

**Bolt group in cover plates:**
*Clause 6.3.4:* Joint length is less than 500 mm therefore no reduction in shear capacity required. Since tension exists M20 H.S.F.G. bolts will be used to eliminate bolt slip.

*Clause 6.4.2.1:* $\quad$ Shear capacity $= P_{SL} = 1.1K_s\mu P_o$
*BS 4604: Part 1* $\qquad P_o = 144$ kN $\quad K_s = 1.0 \quad \mu = 0.55$
$\qquad\qquad$ Shear capacity $= 6 \times (1.1 \times 1.0 \times 0.55 \times 144)$
$\qquad\qquad\qquad\qquad = 475.2$ kN $\geq 128$ kN
**Bolts are adequate in shear**

*Clause 2.4.5.3.(c)* $\quad$ Structural integrity requirements
The column splice should be capable of resisting a tensile force of not less than two-thirds of the factored vertical load applied to the column from the floor level next below the splice, i.e.

Factored vertical load $= 0.67 \times 600$ kN $= 400$ kN (200 kN per cover plate)
$\qquad\qquad\qquad < $ the tension, bearing and shear capacities
**Splice is adequate with respect to structural integrity**

In most cases tension will not occur in the flange plates of a column splice. The most likely location for this to develop is near the top few storeys of a tall frame where the magnitude of the vertical loading is small compared to that at the lower levels and hence the effect of the moment is more significant.

## 5.8 Solution to Example 5.1

| Contract :Connections  Job Ref. No.: Example 5.1 Part of Structure :   Simple Bolted Connections Calc. Sheet No. :  1 of  2 | Calcs. by : W.McK. Checked by : Date : |
|---|---|

| References | Calculations | Output |
|---|---|---|
| | 1 No. 16 mm dia. black bolt / 10 mm / 20 kN / 20 kN / 10 mm / 25 mm / 25 mm / 25 mm / 25 mm | |

<table>
<tr><td>BS:5950:Pt1<br>BCSA</td><td>Structural use of steelwork in building<br>Joints in Simple Construction Volumes 1 & 2</td><td></td></tr>
<tr><td></td><td>Assuming that all edges are either rolled or machine<br>flame cut</td><td></td></tr>
<tr><td>Clause 6.2.3<br>Table 31</td><td>(a) Minimum edge distance $\geq 1.25D$<br>Since 16 mm dia. bolts are being used, the clearance<br>holes are $16 + 2 = 18$ mm dia.<br>$\therefore$ Min. edge/end distance $= \quad 25$ mm<br>$\geq \quad 1.25 \times 18 = 22.5$ mm</td><td>Acceptable</td></tr>
<tr><td>Clause 6.3.2<br><br>Table 32</td><td>(b)<br>(i)<br>Bolt shear capacity $= P_s \quad = p_s A_s$<br>$p_s \quad = 160$ N/mm$^2$<br>Tensile area of 16 mm dia. bolt $\quad = \quad A_t \quad = 157$ N/mm$^2$<br>Bolt shear capacity $= \quad (160 \times 157)/10^3 \quad = 25.1$ kN<br>(ii)</td><td></td></tr>
<tr><td>Clause 6.3.3.2<br>Table 32</td><td>Bolt bearing capacity $\quad\quad P_{bb} = dtp_{bb}$<br>$d = 16$ mm, $t = 10$ mm $\quad\quad P_{bb} = 460$ N/mm$^2$<br>Bolt bearing capacity $= (16 \times 10 \times 460)/10^3 = 73.6$ kN<br>(iii)</td><td></td></tr>
<tr><td>Clause 6.3.3.2<br>Table 33</td><td>Plate bearing capacity $\quad\quad P_{bs} = \quad dtp_{bs}$<br>$\leq 0.5 etp_{bs}$<br>Since in this case $e < 2d$ the bearing capacity will be<br>governed by $0.5 etp_{bs}$<br>$e = 25$ mm, $\quad t = 10$ mm, $\quad\quad p_{bs} = 460$ N/mm$^2$<br>Plate bearing capacity $= (0.5 \times 25 \times 10 \times 460)/10^3 = 57.5$ kN</td><td></td></tr>
<tr><td></td><td>Note: It is normal practice to ensure that the end distance is at<br>least equal to $2d$ and hence the bolt bearing capacity will govern.</td><td>✳✳✳✳</td></tr>
</table>

| Contract :Connections  Job Ref. No. : Example 5.2<br>Part of Structure :   Simple Bolted Connections<br>Calc. Sheet No. :  2   of 2 | Calcs. by : W.McK.<br>Checked by :<br>Date : |
| --- | --- |

| References | Calculations | Output |
| --- | --- | --- |
| *Clause 4.6.1*<br>*Table 33*<br>*Table 6* | (iv)<br>Plate tension capacity    $P_t = A_e p_y$<br>$A_e$  =  $K_e A_{net}$      <  $A_{gross}$<br>$K_e$  =  1.2,     $p_y$  =  275 N/mm$^2$<br>$A_{net}$  =  $(50 - 18) \times 10$  =  320 mm$^2$<br>$A_{gross}$  =  $50 \times 10$  =  500 mm$^2$<br>$A_e$  =  $1.2 \times 320$  =  384 mm$^2$<br><br>Plate tension capacity = $(384 \times 275)/10^3$   = 105.6 kN<br>From (i) to (iv) the capacity of the connection is<br>governed by the shear strength, i.e.<br>Maximum shear which can be transmitted  =  25.2 kN<br>                                  >  20.0 kN | Connection is<br>adequate |

## 5.9 Solution to Example 5.2

| Contract : Connections  Job Ref. No. : Example 5.2<br>Part of Structure :   Simple Bolted Connections<br>Calc. Sheet No. :  1 of  2 | Calcs. by : W.McK.<br>Checked by :<br>Date : |
| --- | --- |

| References | Calculations | Output |
| --- | --- | --- |

4 No. 20 mm dia. black bolts

75 kN

75 kN

150 kN

8 mm

12 mm

8 mm

40 mm

50 mm

40 mm

45 mm   50 mm   45 mm

| References | Calculations | Output |
| --- | --- | --- |
| BS 5950: Pt1<br>BCSA | *Structural use of steelwork in building*<br>*Joints in Simple Construction Volumes 1 & 2*<br><br>Assuming that all edges are either rolled or machine flame cut | |
| Clause 6.2.3<br>Table 31 | (a) Minimum edge distance $\geq 1.25 D$<br>Since 20 mm dia. bolts are being used, the clearance holes are $20 + 2 = 22$ mm dia.<br>$\therefore$ Min. edge/end distance provided $= 40$ mm<br>$\geq 1.25 \times 22 = 27.5$ mm<br>(b) (i)  Bolt shear capacity $P_s = p_s A_s$<br>$p_s = 160$ N/mm$^2$ | Acceptable |
| Clause 6.3.2<br>Table 32 | Tensile area of 20 mm dia. bolt      $A_t = 245$ mm$^2$<br>Bolt shear capacity $=$  $(160 \times 245)/10^3$  $= 39.2$ kN/bolt | |
| Clause 6.3.3.2<br>Table 32 | (ii)  Bolt bearing capacity   $P_{bb} = dt p_{bb}$<br>$d = 16$ mm,  $t = 10$ mm, $P_{bb} = dt p_{bb}$<br>Bolt bearing capacity $= (20 \times 12 \times 460)/10^3 = 110.4$ kN/bolt | |
| Clause 6.3.3.2 | **Note:** the value of $t$ is the lesser of the sum of the outer plate or  the inner plate (i.e. 16 mm or 12 mm) | |
| Table 33 | (iii)  Plate bearing capacity $P_{bs} = \quad dt p_{bs}$<br>$\leq \quad 0.5 e t p_{bs}$<br>$e = 40$ mm, $(\geq 2d)$ $\qquad t = \quad 12$ mm, | ✳ ✳ ✳ ✳ ✳ |

| References | Calculations | Output |
|---|---|---|
| | **Contract :  Connections  Job Ref. No.: Example 5.2** **Calcs. by : W.McK.** **Part of Structure :  Simple Bolted Connections** **Checked by :** **Calc. Sheet No. : 2  of  2** **Date :** | |

| References | Calculations | Output |
|---|---|---|
| | $p_{bs}$ = 460 N/mm$^2$ Plate bearing capacity = $(20 \times 12 \times 460)$ = 110.4 kN/bolt (iv)  Plate tension capacity = $P_t = A_e p_y$ $A_e = K_e A_{net}$  $<$  $A_{gross}$ $K_e = 1.2$,  $p_y = 275$ N/mm$^2$ | |
| *Clause 4.6.1* *Clause 3.3.3* *Table 6* | $A_{net}$     =     $(130 - 44)12$     =     1032  mm$^2$ $A_{gross}$     =     $130 \times 12$     =     1560  mm$^2$ $A_e$     =     $(1.2 \times 1032)/10^3$     =     1238.4  mm$^2$ Plate tension capacity        = $(1238.4 \times 275)/10^3$ = 340.6 kN From (i) to ( iv) the capacity of the connection is governed by the shear capacity, i.e. Maximum. shear which can be transmitted  = $4 \times 39.2$                                                                = 156.8 kN                                                                > 150 kN | Connection is adequate |

## 5.10  Solution to Example 5.5

| References | Calculations | Output |
|---|---|---|

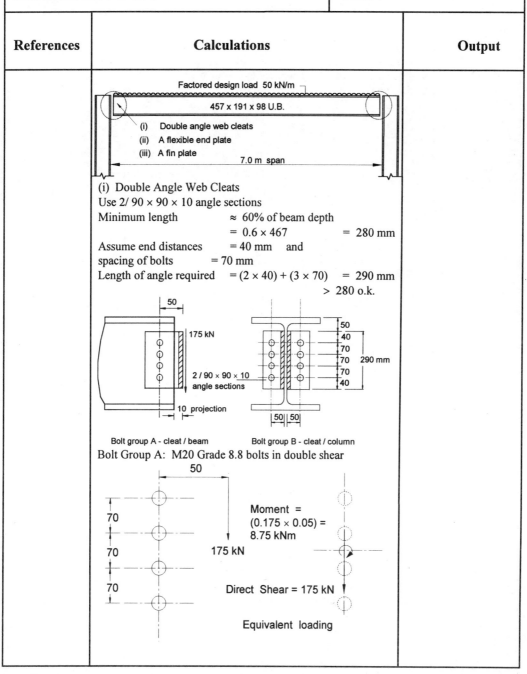

(i)  Double Angle Web Cleats
Use 2/ 90 × 90 × 10 angle sections
Minimum length        ≈ 60% of beam depth
                = 0.6 × 467        = 280 mm
Assume end distances        = 40 mm    and
spacing of bolts        = 70 mm
Length of angle required    = (2 × 40) + (3 × 70)    = 290 mm
                > 280 o.k.

Bolt Group A:  M20 Grade 8.8 bolts in double shear

Moment =
(0.175 × 0.05) =
8.75 kNm

175 kN

Direct Shear = 175 kN

Equivalent loading

| Contract : Connections  Job Ref. No. : Example 5.5<br>Part of Structure :    Simple Bolted Connections<br>Calc. Sheet No. :  2   of   9 | Calcs. by : W.McK.<br>Checked by :<br>Date : |
| --- | --- |

| References | Calculations | Output |
| --- | --- | --- |
| | The equivalent loading on the bolt group results in a 'vertical' shear force acting through the centroid of the bolts in addition to a 'rotational' shear force induced by the moment of 8.75 kNmm.<br><br>Vertical shear force /bolt    $=\dfrac{175}{4}$ kN   =   43.75 kN<br><br>Rotational shear force / bolt   =<br><br>$$\dfrac{\text{Applied Moment} \times \text{Distance to extreme bolt } (y_2)}{\sum \left(\text{Distance from c.of.g. to each bolt}\right)^2}$$<br><br>$$= \dfrac{8.75 \times 10^3 \times 105}{\left(2 \times 35^2\right) + \left(2 \times 105^2\right)} = 37.5 \text{ kN}$$<br><br>43.75 kN        Resultant shear force<br>in the top bolt<br><br>Resultant maximum shear force    $= \sqrt{43.75^2 \; + \; 37.5^2}$<br>                                                      $= 57.6$ kN | |
| Clause 3.6.2 | Double shear capacity of 4 Nº M20 Grade 8.8 bolts<br>$P_s = 2(p_sA_s) = 2(375 \times 245) / 10^3$   $= 183.75$ kN<br>                                    $P_s \; > \; 57.6$ kN | Bolts adequate in shear |

| Contract : Connections  Job Ref. No. : Example 5.5<br>Part of Structure :   Simple Bolted Connections<br>Calc. Sheet No. :  3 of 9 | Calcs. by : W.McK.<br>Checked by :<br>Date : |
| --- | --- |

| References | Calculations | Output |
| --- | --- | --- |
| (b) Bolt Bearing<br><br><br><br><br><br><br><br><br><br><br><br><br><br><br><br><br><br><br><br><br><br><br>*Clause 6.3.3.3*<br><br><br><br><br><br>*Table 33* | 4 N° M20 Grade 8.8 bolts in 10 mm thick plate<br>$P_b = 4(970 \times 10 \times 20) / 10^3 = 776$ kN<br>Cleat shear<br><br><br><br>shear force =    87.5 kN / cleat<br><br>Shear Capacity =    $P_v = 0.6p_yA_v$<br>where:<br>$A_v = 0.9[(290 \times 10) - (4 \times 22 \times 10)]$    =    1818 mm²<br>$P_v = (0.6 \times 275 \times 1818)/10^3$         =    300 kN<br>                                    $P_v$  >    87.5 kN<br><br>Cleat Bearing<br>Bearing Capacity        $P_{bs}$  =  $dtp_{bs}$   ≤   $0.5etp_{bs}$<br>$e = 40$  since $e = 2d$      $dtp_{bs} = 0.5etp_{bs}$<br>$p_{bs} = 460$ N/mm²<br>$P_{bs} = (20 \times 10 \times 460)/10^3 = 92$ kN     >    57.6/2 = 28.8 kN<br><br>Beam Web Shear<br><br> | <br><br><br><br><br><br><br><br><br><br><br><br><br><br><br><br><br><br><br><br>Cleat adequate in<br>shear<br><br><br><br><br>Cleat is adequate<br>bearing |

| | | |
|---|---|---|
| **Contract : Connections  Job Ref. No. : Example 5.5** **Part of Structure :   Simple Bolted Connections** **Calc. Sheet No. :  4  of  9** | | **Calcs. by : W.McK.** **Checked by :** **Date :** |

| References | Calculations | Output |
|---|---|---|
| *Clause 4.2.3* *Table 6* | Shear Capacity $P_v$ = $0.6 p_y A_v$ Assume   $A_v$ = $0.9 [ (290 \times 11.4) - (4 \times 22 \times 11.4) ]$ = $2073$ mm$^2$ $P_v \approx (0.6 \times 275 \times 2073)/10^3$ = $342$ kN $P_v >$ $175$ kN | Beam web is adequate in shear |
| *Clause 6.3.3.3* | Beam Web Bearing As before   $e = 2d = 40$ mm $P_{bs}$ = $(20 \times 11.4 \times 460)/10^3$ = $104.8$ kN $P_{bs} >$ $57.6$ kN Bolt Group B : M20 Grade 8.8 bolts in single shear. | Beam web is adequate in bearing |
| | Shear force = 175 kN | |
| *Clause 6.3.2* *Table 32* | Shear Force / bolt  = 175 / 8  = 21.9 kN $P_s = (375 \times 245)/10^3 = 91.9$ kN $P_s > 21.9$ kN | Bolts are adequate in shear |
| *Clause 6.3.3.2* *Table 33* | Bolt Bearing $P_{bb} = dt p_{bb}$   where $p_{bb} = 460$ N/mm$^2$ $P_{bb} = (20 \times 10 \times 460)/10^3 = 92$ kN $P_{bb} > 21.9$ kN | Bolts are adequate in bearing |
| *Clause 4.2.3* *Table 6* | Cleat Shear Shear Force/ Cleat  = 175 / 2 = 87.5 kN $P_v = 0.6 p_y A_v$ $p_y = 275$ N/mm$^2$ ;      $A_v$      = $1818$ mm$^2$ $P_v = (0.6 \times 275 \times 1818)/10^3$  = $300$ kN | Cleats are adequate in shear |
| *Clause 6.3.3.3* | Cleat Bearing $P_{bs} = dt p_{bs} \leq 0.5 e t p_{bs}$ | |

| References | Calculations | Output |
|---|---|---|
| | **Contract : Connections  Job Ref. No. : Example 5.5**<br>**Part of Structure :   Simple Bolted Connections**<br>**Calc. Sheet No. :  5  of 9** | **Calcs. by : W.McK.**<br>**Checked by :**<br>**Date :** |

| References | Calculations | Output |
|---|---|---|
| *Table 33* | As before $e = 2d$<br>$P_{bs} = (20 \times 10 \times 460)/10^3 = 92$ kN<br>$P_{bs} > 21.9$ kN | Cleats are adequate in bearing |
| | (b) Flexible End Plate<br>Using a 150 × 8 mm thick end plate<br>Minimum length $\approx 0.6D$  i.e. 290 mm as before | |
| | | |
| *Clause 6.3.2*<br>*Table 32* | Bolt Shear<br>Shear capacity in single shear $P_s$   = 91.9 kN<br>Maximum design shear force/bolt   = 175 / 8 = 21.75 kN<br>$\qquad\qquad\qquad\qquad P_s > 21.75$ kN | Bolts are adequate in shear |
| *Clause 6.6.3.2* | Bolt Bearing<br>Bolt bearing capacity   $P_{bb} = (20 \times 8 \times 460) / 10^3$<br>$\qquad\qquad = 73.6$ kN   >   21.75 kN | Bolts are adequate in bearing |
| *Clause 4.2.3* | Plate shear<br> | |
| *Table 6* | $\qquad$ shear = 175 kN shear / section  =   87.5 kN<br>$P_v = 0.6p_y A_v$<br>$A_v = 0.9 [(290 \times 8) - (4 \times 22 \times 8)]$   =   1454 mm²<br>$P_v = (0.6 \times 275 \times 1454) / 10^3$   =   240 kN<br>$\qquad\qquad\qquad P_v >$   87.5 kN | Plate adequate in shear |

| References | Calculations | Output |
|---|---|---|
| *Clause 6.3.3.3*<br>*Table 33* | Plate bearing<br>$P_{bs} = dtp_{bs} \leq 0.5etp_{bs}$        $e = 2d$<br>$P_{bs} = (20 \times 8 \times 460) / 10^3 = 73.6 \text{ kN}$    $> 21.75 \text{ kN}$ | Plate adequate in bearing |
| | Web shear<br> | |
| *Clause 6.6.2* | $P_v = 0.6p_yA_v$<br>$A_v = (0.9 \times 290 \times 11.4)$        $= 2975 \text{ mm}^2$<br>$P_v = (0.6 \times 275 \times 2975) / 10^3$    $= 491 \text{ kN}$    $> 175 \text{ kN}$<br><br>Weld<br>Assuming 6 mm fillet welds, E43 electrodes | Beam web is adequate in shear |
| *Clause 6.6.5*<br>*Table 36* | Effective length of weld $= 2[290 - (2 \times 6)]$    $= 556 \text{ mm}$<br>$p_w = 215 \text{ N/mm}^2$   throat size $= 0.7 \times 6$    $= 4.2 \text{ mm}$<br>Strength of weld / mm $= (4.2 \times 215) / 10^3$    $= 0.903 \text{ kN}$<br>Weld capacity $= 0.903 \times 556 = 502 \text{ kN}$    $> 175 \text{ kN}$ | 556 mm length 6 mm fillet weld is adequate |
| | (c) Fin Plates<br> | |

| Contract :Connections  Job Ref. No. : Example 5.5<br>Part of Structure :    Simple Bolted Connections<br>Calc. Sheet No. :  7  of  9 | Calcs. by : W.McK.<br>Checked by :<br>Date : |
|---|---|

| References | Calculations | Output |
|---|---|---|
| | Fin plate thickness  $t =$  10 mm  $\leq 0.5d$<br>Fin plate length     $l \leq$  0.6 × 467 = 280 mm   say 290 mm<br>Applied vertical shear force = 175 kN<br><br>(a) Bolt bearing<br>$t_{web} = 11.4$ mm,          $t_{\text{fin plate}}$ = 10 mm<br><br> | |

$$P_{bs} = dtp_{bs} \quad = \quad \frac{20 \times 10 \times 460}{10^3} = 92 \text{ kN}$$

Vertical shear force / bolt       $= \dfrac{175}{4} = $ 43.75 kN

Moment    =    $(175 \times 60)/10^3$ =    10.5 kNm

Rotational shear force / bolt    $= \dfrac{10.5 \times 10^3 \times 105}{2\left(35^2 + 105^2\right)}$

$$= 45 \text{ kN}$$

Resultant shear force on outermost bolt

$$= \sqrt{43.75^2 + 45^2}$$
$$= 62.8 \text{ kN}$$
$$P_{bs} > 62.8 \text{ kN}$$

(b) Bolt shear
$P_s \; = p_s A_s \; = (375 \times 245) / 10^3 = 91.9$ kN
$$P_s \; > 55.4 \text{ kN}$$

(c) Beam web shear
As for beam web shear with double angle web cleats
(see page 189 )

Block shear:

$e_1 \; \leq 2.5d \; = \; 2.5 \times 20 \; = \; 50$ mm
    $\leq \; e_t \; = \; 50 + 40 \; = \; 90$ mm      $\therefore e_1 = 50$ mm

References column: *Table 33*

Output column: Web and Plate are adequate in bearing ; Bolts are adequate in in bearing and shear

| Contract : Connections  Job Ref. No. : Example 5.5<br>Part of Structure :    Simple Bolted Connections<br>Calc. Sheet No. : 8 of 9 | Calcs. by : W.McK.<br>Checked by :<br>Date : |
| --- | --- |

| References | Calculations | Output |
| --- | --- | --- |
| | $e_2 \leq 2.5d$   =   $2.5 \times 20$   =   50 mm<br>   $\leq e_b$   $\geq$   50  By inspection             $\therefore e_2 = 50$ mm<br><br>$D_{net}$   =   $(210 + 50 + 50) - (4 \times 22)$ =   222 mm<br><br>Block shear capacity   $\approx$   $0.6p_yA_v$<br><br>$A_v = D_{net} \times t_w = 222 \times 11.4$    = 2530.8 mm<br>$P_{vb} = (0.6 \times 275 \times 2530.8)/10^3 = 417.6$ kN > 175 kN<br><br>(d) Fin plate shear<br>The fin plate is 10 mm thick and therefore has a slightly lower shear capacity than the 11.4 mm thick web. By inspection, this is still adequate. | Adequate with respect to block shear |
| *Appendix B*<br><br>*Clause B.2.7* | (e) Fin Plate Bending<br><br>$M_b = S_{xx}p_b$<br><br>$S_{xx} = \dfrac{tl^2}{4} = \dfrac{10 \times 290^2}{4} = 210.25 \times 10^3 \, mm^3$<br><br>$\lambda_{LT} = n \times 2.8 \times \left(\dfrac{L_E d}{t^2}\right)^{\frac{1}{2}}$<br><br>where *n* is as given in *Clause 4.3.7.6* and equals 0.77<br><br>$L_E = 60$ mm<br><br>175 kN<br>290 mm<br>60  40<br><br>$\lambda_{LT} = 0.77 \times 2.8 \times \left(\dfrac{60 \times 290}{10^2}\right)^{\frac{1}{2}} = 28.4$ | |

| | | |
|---|---|---|
| **Contract : Connections  Job Ref. No. : Example 5.5** | **Calcs. by : W.McK.** | |
| **Part of Structure :   Simple Bolted Connections** | **Checked by :** | |
| **Calc. Sheet No. :  9  of  9** | **Date :** | |

| References | Calculations | Output |
|---|---|---|
| *Table 12* | $p_y$ = 275 N/mm$^2$      $\lambda_{LT}$ = 28.4      $\therefore$ $p_b$ = 275 N/mm$^2$ <br><br> $M_b$ = $(210.25 \times 10^3 \times 275)/10^6$   =   57.8 kNm <br><br> $M_x$ = $(175 \times 60)/10^3$                 =   10.5 kNm <br> $\qquad\qquad\qquad\qquad$ $M_b$    >    $M_x$ <br><br> Weld between fin plate and column <br><br> <br> 290mm <br> 1.0⊣ ⊢⊣ ⊢1.0 <br><br> Consider  a pair of welds each of unit width × 290 mm long <br><br> Applied moment        = 10.5 kNm <br> Applied shear force= 175 kN <br> Total length of weld    = 2 × 290                = 580 mm <br> Direct vertical shear force/mm =   175/580 = 0.3 kN/mm <br><br> $$I_{xx} = 2 \times \left(\frac{1 \times 290^3}{12}\right)^4 = 4.06 \times 10^6 \text{ mm}^4$$ <br> Distance to the extreme fibre = 145 mm <br><br> Max. bending force/mm = $\dfrac{\left(10.5 \times 10^3\right) \times 145}{4.06 \times 10^6}$ = 0.38 kN/m <br><br> Resultant Force/mm = $\sqrt{0.3^2 + 0.375^2}$ = 0.48 kN / mm <br> Strength of 8 mm fillet weld = 1.2 kN/mm <br> $\qquad\qquad\qquad$ $\geq 0.8 \times$ plate thickness | Fin plates are adequate in bending <br><br><br><br><br><br><br><br><br><br><br><br><br><br><br><br><br><br><br> Adopt 8 mm fillet welds |

## 5.11  Solution to Example 5.6

| Contract : Connections  Job Ref. No. : Example 5.6<br>Part of Structure :   Bolted Moment Connections<br>Calc. Sheet No. :  1  of  2 | Calcs. by : W.McK.<br>Checked by :<br>Date : |
| --- | --- |

| References | Calculations | Output |
| --- | --- | --- |
| | <br>Flange thickness = 19.7mm<br>609.6mm<br><br>Ultimate design moment     =    540 kNm<br>Ultimate design shear      =    380 kN<br>Assume that the moment is transferred by the flange<br>bolts whilst the shear is transferred by the web bolts.<br><br>Flange force induced by the moment = $F_t = \dfrac{Moment}{lever - arm}$<br><br>$F_t = \dfrac{540 \times 10^6}{(609.6 - 19.7)}$  = 915.4 kN<br><br>Moment:<br>Tension capacity  $P_t = 0.9P_o \geq \dfrac{915.4}{4}$   =   229 kN | |
| *Clause 6.4.4.2* | Min.   $P_o$ required  $= \dfrac{229.0}{0.9}$  = 254.4 kN<br>For  30 mm H.S.F.G  Bolts   $P_o$ = 286 kN | Adopt 4 No. 30 mm<br>H.S.F.G.  Bolts |
| | Shear:<br>Slip resistance  $P_{SL} = 4 \times (1.1 \times K_s \mu P_o)$      $\geq$ 380 kN<br>where  $K_s = 1.0$  and  $\mu = 0.45$<br>$P_{SL \; provided} = 4 \times 1.1 \times 0.45 \times 286 = 566$ kN  $\geq$ 380 kN | Bolts are adequate<br>in shear |
| *Clause 6.4.2.1* | Bearing resistance  $P_{BG} = 4 \times (dtp_{bg})$      $\geq$ 380 kN<br>$P_{BG \; provided} = 4 \times (30 \times 12 \times 825) / 10^3$      = 1188 kN<br>      $\geq$ 380 kN | Bolts are adequate<br>in bearing |

| References | Calculations | Output |
|---|---|---|
| *Clause 6.4.2.2* | Clearly in this case the shear and bearing capacity of the connection are more than adequate.<br><br>The welds, end-plate and stiffeners must also be designed to transfer the appropriate loads and in addition the column flange and web should be checked.<br>The end-plates, and column web and flanges and the welds must also be designed to transfer the appropriate load. If necessary additional strengthening such as stiffeners may be required to enhance the buckling or bearing capacity of the column web.<br>The detailed design required for this is beyond the scope of this publication. A rigorous, detailed analysis and design procedure for connections and their component parts can be found in *Joints In Steel Construction : Moment Connections*, published by The Steel Construction Institute. | |

**Contract : Connections  Job Ref. No. : Example 5.6**
**Part of Structure :   Simple Bolted Connections**
**Calc. Sheet No. :  2  of  2**

**Calcs. by : W.McK.**
**Checked by :**
**Date :**

# 6. Plate Girders

## 6.1 Introduction

In many instances, it is necessary to support heavy vertical loads over long spans resulting in relatively large bending moments and shear forces. If the magnitude of these is such that the largest Universal Beam section, even when compounded with plates, is inadequate then it is necessary to fabricate a beam utilizing plates welded together into an I-shaped section.

The primary purpose of the flange plates is to resist the tensile and compressive forces induced by the bending moment. The primary purpose of the web plate is to resist the shearing forces. These sections are normally more efficient in terms of steel weight than rolled sections, particularly when variable depth girders are used, since they can be custom designed to suit requirements. In addition, the development of automated workshops has reduced fabrication costs, particularly when compared to box girders and trusses which are still fabricated manually resulting in high costs.

When compared to other forms of construction, plate girders do have some disadvantages, e.g. they are heavier than comparable trusses, more difficult to transport, have low torsional stiffness, can be susceptible to instability of the compression flange during erection and do not easily accommodate openings for services.

Since plate girders are normally highly stressed, the possibility of local buckling will tend to be higher than in rolled sections, particularly with respect to the web plate. In most cases, stiffeners are provided along the length of the girder to prevent the web plate from buckling.

A typical plate-girder and its component parts is shown in Figures 6.1 and 6.2; in addition a flow chart is shown in Figure 6.3, indicating the principal steps which are undertaken when designing such a girder.

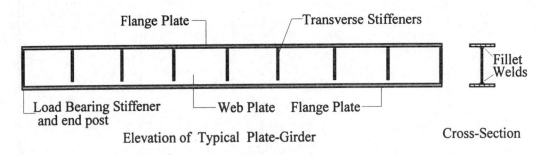

Elevation of Typical Plate-Girder

Cross-Section

**Figure 6.1**

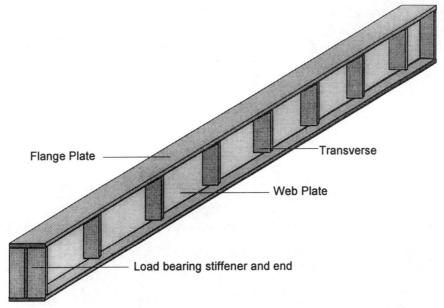

**Figure 6.2** Typical plate girder

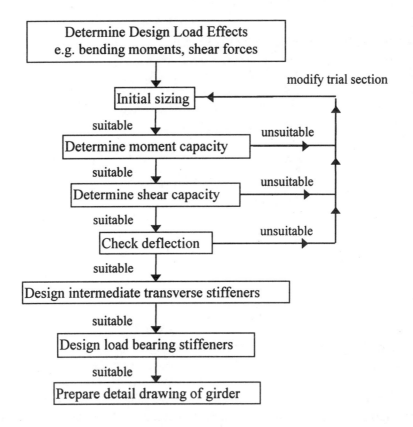

**Figure 6.3**

### *6.1.1 Design Load Effects*

The design load effects on a plate-girder are determined using standard structural analysis techniques. They comprise the effects induced by bending which are primarily,

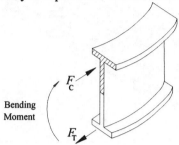

**Figure 6.4**

(i) Compressive and tensile forces above and below the neutral axis of the cross-section

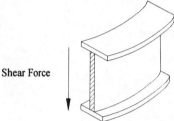

**Figure 6.5**

(ii) Shearing forces which are resisted by the web

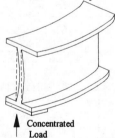

**Figure 6.6**

(iii) Forces tending to induce failure by web buckling

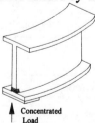

**Figure 6.7**

(iv) Forces tending to induce failure by web crushing
(v) Vertical deflection

The magnitudes and senses of these effects are dependent on the magnitude and nature of the forces applied to the girder. The service design loading for buildings is given in BS 6399:Part 1 and the appropriate load factors can be found in *Table 2* of BS 5950:Part 1.

## 6.2 Initial Sizing

The initial sizing of the component parts of a plate girder is carried out on the basis of experience and consideration of the section classification as given in BS 5950:Part 1. While webs tend to be slender, they should not be excessively so, since additional stiffening will be required with its inherent fabrication costs.

Tables 6.1 and 6.2 given below indicates typical values of span-to-depth ratios and web thicknesses for plate girders used in buildings.

**Table 6.1**

| | Applications | Typical span-to-depth ratio |
|---|---|---|
| (i) | simply-supported, non-composite girders with concrete decking; constant depth beams used in simply-supported composite girders | 12 - 20 |
| (ii) | constant depth beams used in continuous non-composite girders using concrete decking | 15 - 20 |
| (iii) | simply-supported crane girders | 10 - 5 |

**Table 6.2**

| | Beam depth (mm) | Typical thickness of web (mm) |
|---|---|---|
| (a) | Up to 1200 | 10 |
| (b) | 1200 - 1800 | 12 |
| (c) | 1800 - 2250 | 15 |
| (d) | 2250 - 3000 | 20 |

## 6.3 Moment and Shear Capacity

The moment capacity of plate girders where the flanges of the section are plastic, compact or semi-compact, but where the web is slender (as defined in *Table 7*) or is thin and susceptible to shear buckling (i.e. $d/t \geq 63\varepsilon$), can be evaluated using one of three methods outlined in *Clause 4.4.4.2* of BS 5950:Part 1. Method (a) in which the moment is assumed to be resisted by the flange plates alone subjected to a uniform stress of $p_y$ and the shear stress is assumed to be resisted by the web is adopted in this text. This method is considerably easier to understand and use than the other two that are given.

The implication in the code of using a uniform stress equal to $p_y$, is that the compression flange is fully restrained. In instances where this does not occur, a value of $p_b$ can be determined using the equivalent slenderness $\lambda_{LT}$ and *Table 12*. The equivalent slenderness can be evaluated using *Clause 4.3.7* or the equations given in *Appendix B* of the code.

For the design of sections in which the flanges are slender, a reduction factor should be applied to the uniform stress value; this can be found in *Table 8* of the code.

The section classification of the web as indicated in section 2.2 relates to local buckling induced by longitudinal compressive stresses. Webs can also fail due to local buckling caused by concentrated loads applied to the flanges, as indicated in Section 2.5.1. In addition to these two modes of failure, a thin web can fail owing to shear buckling in which diagonal compressive and tensile stresses are developed within web panels between stiffeners. The compressive stresses induce out-of-plane shear buckling and tension-field action subsequently enhances the load-carrying capacity until failure occurs with the development of plastic hinges in the flanges, as shown in Figures 6.8 and 6.9.

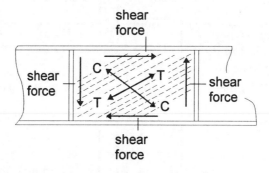

**Figure 6.8**

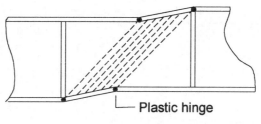

**Figure 6.9**

The code permits allowance for tension-field action in the design of webs and stiffeners. This technique enables a more accurate assessment of the contribution made by stiffeners to the strength of the web and hence a more economic section to be designed. Tension-field action is not considered in this text.

The shear capacity of a section when tension-field action is not being considered is given in *Clause 4.4.5.3*. This is dependent on the critical shear stress $q_{cr}$, given in *Table 21* and will determine if intermediate stiffeners are required.

## 6.4 Deflection

As with Universal Beam design, the deflection induced by the unfactored imposed loads should not exceed the limits specified in *Table 5* of BS 5950:Part 1. The actual deflection can be calculated using coefficients or the equivalent uniform load distribution technique, both of which are illustrated in the example at the end of this chapter.

## 6.5 Intermediate Stiffeners

Since, for economy, plate girders are designed with thin webs (i.e. *Clause 3.6.2* BS 5950:Part 1, $d/t \geq 63\varepsilon$), shear buckling must be considered. The serviceability requirements which require a minimum thickness to prevent buckling of the web and of the flange into the web are given in *Clauses 4.4.2.2* and *4.4.2.3* of BS 5950:Part 1 respectively. As mentioned previously, the shear buckling resistance of webs can be calculated from *Clause 4.4.5.3*. If this value is exceeded, then it will be necessary to include intermediate stiffeners to prevent out-of-plane buckling of the web.

## 6.6 Load Bearing Stiffeners

When a concentrated load is applied through the flange, for example at a point of support or under a column, it is sometimes necessary to provide load-bearing stiffeners, which may be single or double, to prevent local buckling and crushing of the web. The load carrying capacity of the web in buckling and bearing can be determined using *Clauses 4.5.2* and *4.5.3* (see Chapter 2 Section 2.5 for universal beam design).

Where load-bearing stiffeners are provided at supports they are also known as *end posts*. The design of end posts including their connection to the web and flange plates is set out in *Clauses 4.5.4* to *4.5.11*. In some instances it is necessary to check the compression edge of the web between stiffeners; the procedure for this is given in *Clause 4.5.2.2* and is illustrated in Example 6.2 of this chapter.

## 6.7 Example 6.1 Plate girder in multi-storey office block

A four-storey office development is to be designed such that approximately 108 m$^2$ of office space is provided in the upper floors, and only perimeter columns are to be used on the ground floor to accommodate parking facilities. Design a suitable plate girder to satisfy the structural arrangement indicated in Figures. 6.10 (a) and (b).

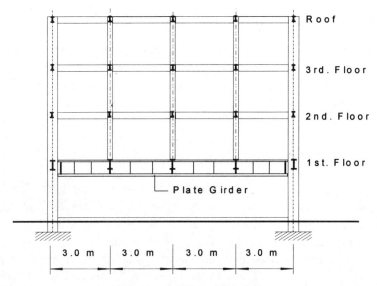

**Figure 6.10(a)** Elevation

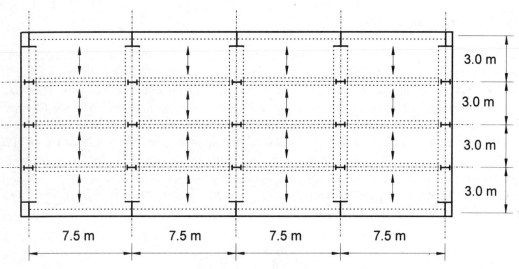

**Figure 6.10(b)** Plan

**Design Data:**

| | |
|---|---|
| Characteristic dead load on the roof (including. self-weight) | = 4.0 kN/m$^2$ |
| Characteristic dead load on the floors (including. self-weight) | = 5.0 kN/m$^2$ |
| Estimated self-weight of the plate girder | = 4.5 kN/m |

Characteristic imposed loads on the roof and floors as per BS 6399:Part 1

Assume that the compression flange is fully restrained throughout and that *Grade 43* steel is used.

### 6.7.1 Design Loading

The design loading on the girder is determined using static analysis in which the floor and roof beams are assumed to be simply supported. This results in the girder being subjected to a uniformly distributed load due to the self-weight, and three concentrated loads at the internal column positions.

The code of practice for the dead and imposed loading on buildings (BS 6399:Part 1) permits a reduction in the value of the imposed loads on multi-storey structures. In cases where a single span of a beam or girder supports not less than $40m^2$ of floor area at one general level, the reduction is in accordance with *Table 3*.

In this design the area of floor supported is equal to (7.5 m × 12.0 m) = 90 $m^2$ and therefore the imposed load should be reduced accordingly, i.e. using linear interpolation and the values given in *Table 3* the reduction equals 6.25%. It should be noted that the *Table 3* values do not apply to roof loadings (*Clause 5.2*).

### 6.7.2 Column Loads

A typical internal column supports the roof/floor area shown in Figure 6.11, i.e. $(7.5 \times 3.0) = 22.5$ $m^2$

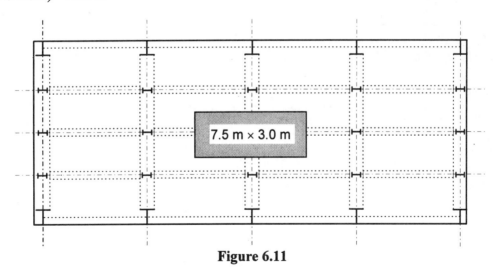

7.5 m × 3.0 m

**Figure 6.11**

| | | | |
|---|---|---|---|
| Service dead load on the roof | = | 4.0 × 22.5 | = 90 kN |
| Service imposed load on the roof (BS 6399:Part 1, *Clause 6.0*) | | | = 1.5 × 22.5 |
| | | | = 33.75 kN |
| Service dead load on each floor | = | 5.0 × 22.5 | = 112.5 kN |

Service imposed load on each floor (BS 6399:Part 1, *Clause 6.0*) allowing for a 6.25% reduction in floor loadings = (1− 0.0625) × (2.5 × 22.5) = 52.73 kN

Design load / column =

$$1.4\left[90 + (3 \times 112.5)\right] + 1.6\left[33.75 + (3 \times 52.73)\right] \qquad = \ 905.6 \ \text{kN}$$

Estimated design self-weight of plate-girder $\qquad = \ 1.4 \times 4.5 \quad = \ 6.3 \ \text{kN/m}$

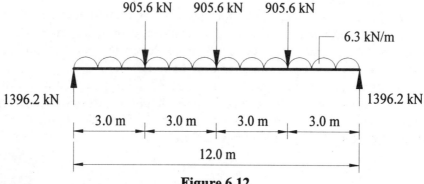

Figure 6.12

Design bending moment $\ = \ (1396.2 \times 6) - (905.6 \times 3) - (6.3 \times 6.0 \times 3)$
$\qquad\qquad\qquad\qquad\quad = \ 5547 \ \text{kNm}$
Design shear force $\qquad = \ 1396.2 \ \text{kN}$

### 6.7.3  Initial Sizing

Assume a span/depth ratio of 12, (see Table 6.1),

The depth of the girder : $d \ \approx \ \dfrac{1200}{12} = \ 1000 \ \text{mm}$

If the flange plate thickness is assumed to be greater than 40 mm then the design strength $p_y$ from BS 5950:Part 1 *Table 6* is 255 N/mm$^2$

Using method (a) mentioned previously, in which the moment capacity of the section is based on the flanges alone as shown below:

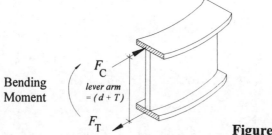

Bending
Moment

$F_C$

lever arm
$= (d + T)$

$F_T$

Figure 6.13

an estimate of the flange force and hence the flange area required can be made.

Flange force $F_c = F_t \approx \dfrac{\text{Bending Moment}}{\text{depth } d} = \dfrac{5547 \times 10^3}{1000} = 5547 \text{ kN}$

$F_c = \text{Area of the flange} \times \text{stress in the flange}$

Area of flange required $\approx \dfrac{5547 \times 10^3}{255} = 21753 \text{ mm}^2$

A flange plate 450 mm wide × 50 mm thick provides an area of 22500 mm². Assume a 10 mm thick web plate (see Table 6.2). The trial girder cross-section is as shown in Figure 6.14.

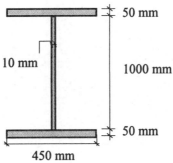

**Figure 6.14**

The cross-sectional area of the proposed girder $= 2 \times (450 \times 50) + (1000 \times 10)$
$= 55 \times 10^3 \text{ mm}^2$

Unit weight of steel $= 7.85 \times 10^{-8} \text{ kN/mm}^3$

Self-weight of the girder / m $= \left(55 \times 10^3\right) \times \left(1000\right) \times \left(\dfrac{7.85}{10^8}\right) = 4.32 \text{ kN/m}$

This is slightly less than the self-weight assumed in the initial sizing of the girder.

### 6.7.4 Section Classification    (Clause 3.5)

The local buckling of cross-sections can be avoided by limiting the width to thickness ratios of individual elements which are subjected to compression. The various elements of a cross-section, predominantly the flanges and webs, are classified in BS 5950:Part 1, *Table 7* as being plastic, compact, semi-compact or slender, the definitions of which are given in *Clause 3.5.2*. Where elements of a cross-section are shown to be slender, then additional measures which must be taken to prevent local buckling are given in *Clause 3.6*.

### 6.7.5 Flanges

$T = 50 \text{ mm}$   from *Table 6*   $p_y = 255 \text{ N/mm}^2$

$$b = \frac{(B-t)}{2} = \frac{(450 - 10)}{2} = 220 \text{ mm}$$

$$\frac{b}{T} = \frac{220}{50} = 4.4; \qquad \varepsilon = \sqrt{\frac{275}{p_y}} = \sqrt{\frac{275}{255}} = 1.04$$

In *Table 7* flanges in which $\dfrac{b}{T} \le 7.5\varepsilon$ are considered to be plastic.

### 6.7.6  Web

$t = 10$ mm  from *Table 6* $\qquad p_{yw} = 275$ N/mm$^2$

$$\frac{d}{t} = \frac{1200}{10} = 120 ; \qquad \varepsilon = \sqrt{\frac{275}{275}} = 1.0$$

In *Table 7* webs in which $98\varepsilon \ < \ \dfrac{d}{t} \ \le 120\varepsilon$ are considered to be semi-compact

*Clause 3.6.2*

In addition, since $\dfrac{d}{t} > 63\varepsilon,$ then *Clause 4.4.4.2* should be used to determine the capacity

of the section since the web is thin and is required to carry shear (i.e. *method (a)* described previously).

*Clause 4.4.2.2*

To ensure that the web will not buckle under normal service conditions it is necessary to satisfy the following minimum web thickness criteria:

(a)  without intermediate stiffeners : $\qquad t \ge \dfrac{d}{250}$

(b)  with transverse stiffeners only :

    (i)    where stiffener spacing $\qquad a > d$ then $t \ge \dfrac{d}{250}$

    (ii)   where stiffener spacing $\qquad a \le d$ then $t \ge \dfrac{d}{250}\left(\dfrac{a}{d}\right)^{\frac{1}{2}}$

Assuming the more critical case with stiffeners, then $\quad t \ge \dfrac{1200}{250} = 4.8$ mm

*Clause 4.4.2.3*

Similarly, to ensure that the flange does not buckle into the web, it is necessary to satisfy the following criteria:

(a)  without intermediate stiffeners: $t \ \ge \ \dfrac{d}{250}\left(\dfrac{p_{yf}}{345}\right)$

(b)  with intermediate transverse stiffeners:

(i) where stiffener spacing $a > 1.5d$ then $t \geq \dfrac{d}{250}\left(\dfrac{p_{yf}}{345}\right)$

(ii) where stiffener spacing $a \leq 1.5d$ then $t \geq \dfrac{d}{250}\left(\dfrac{p_{yf}}{455}\right)^{\frac{1}{2}}$

Assuming the more critical case with stiffeners, i.e. $a \geq 1.5d$,

$$\text{then } t \geq \frac{1000}{250}\left(\frac{255}{345}\right) = 3.0 \text{ mm}$$

Clearly the web is satisfactory in both respects.

### 6.7.7 Moment Capacity    (Clause 4.4.4.2 (a))

The moment carrying capacity of the section ignoring the web is given by :

$$M = p_{yf} \times S_{xx}$$

where:

$S_{xx}$    is the plastic section modulus based on the flanges alone and equals (flange area × distance between the centroids of the flanges)

$S_{xx}$   = $(450 \times 50) \times (1050)$   = $23.625 \times 10^6 \text{ mm}^3$

$M$   = $255 \times 23.625$       = $6024$ kNm.

$M_{\text{applied}} < M$   therefore the section is adequate with respect to bending

**Note:** If the compression flange had not been fully restrained, then the value of $p_{yf}$ would have been replaced by a value of $p_b$ from *Table 12*. This is considered at the end of this chapter in Example 6.2.

### 6.7.8 Shear Capacity    (Clause 4.4.5.3)

Designing *without* tension-field action, the shear capacity of the section is given by:

$$V_{cr} = q_{cr}\,d\,t \quad \geq \quad \text{design shear force}$$

where $q_{cr}$ is the critical shear strength obtained from *Tables 21(a)* to *(d)* as appropriate. If no intermediate stiffeners are assumed the spacing is taken as infinity.
The above equation can be rearranged such that

minimum $q_{cr}$ required    $\geq \dfrac{\text{design shear force}}{dt}$

$$q_{cr} \geq \frac{1396.2 \times 10^3}{1000 \times 10} = 139.6 \text{ N/mm}^2$$

An extract from *Table 21 (b)* is given in Figure 6.15.
It is evident that when $d/t = 120$ and $p_y = 275$ N/mm$^2$ a value of $a/d < 0.78$ will ensure
a value of $q_{cr} \geq 139.6$ N/mm$^2$

| | Stiffener Spacing ratio $a/d$ | | | | | |
|---|---|---|---|---|---|---|
| $d/t$ | 0.4 | 0.5 | 0.6 | **0.7** | **0.8** | 0.9 |
| - | - | - | - | - | - | - |
| - | - | - | - | - | - | - |
| 115 | 165 | 165 | 165 | 154 | 142 | 132 |
| **120** | 165 | 165 | 162 | **149** | **137** | 126 |
| 125 | 165 | 165 | 158 | 144 | 131 | 120 |
| - | - | - | - | - | - | - |

**Figure 6.15**  Extract from *Table 21(b)*

The stiffener spacing  $a \leq 0.78 \times 1000 = 780$ mm;
Adopt a value of spacing $= 750$ mm
A suitable arrangement of stiffeners is shown in Figure 6.16

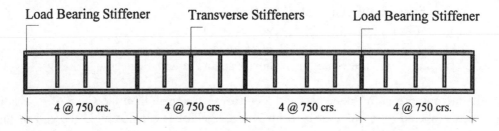

| Load Bearing Stiffener | Transverse Stiffeners | | Load Bearing Stiffener |

| 4 @ 750 crs. | 4 @ 750 crs. | 4 @ 750 crs. | 4 @ 750 crs. |

**Figure 6.16**

In the case of a girder supporting a larger uniformly distributed load the shear force may
reduce significantly towards the centre and hence permit an increase in the stiffener
spacing towards the middle of the span.

### 6.7.9 Deflection  (Clause 2.5)

*Clause 2.5.1*
In *Table 5* the deflection limit for beams without plaster or brittle finishes is given as

$$\delta_{max} \leq \frac{span}{200}$$

This is the deflection induced by the *unfactored imposed load* only, i.e. the service
imposed load.

The actual deflection can be determined using appropriate coefficients such as those given in Table 2.1 in Chapter 2, or can be *estimated* using the equivalent uniform load technique.

*Clause 3.1.2*
*Table 6* the modulus of elasticity      $E$ =   205 kN/mm$^2$
Service load / column                         =   $(33.75 + (3 \times 52.73)) = 191.9$ kN
   The mid-span deflection for a beam with three point loads can be evaluated by super-imposing the contributions from each load i.e.

$$\delta_{max} \approx 2 \left\{ \frac{191.9 \times (12000)^3}{48EI} \left[ 3 \times 3 - 4 \left( \frac{3}{12} \right)^3 \right] \right\} + \frac{191.9 \times (12000)^3}{48EI}$$

$\delta_{max} \approx 6.43$ mm

**Note:** The 2 above can be used in this instance because of the symmetry.

$$Table\ 5\ value \quad = \quad \frac{12000}{200} \quad = \quad 60 \text{ mm}$$

(b) Equivalent U.D.L. technique
Loading considered for deflection calculation is as shown in Figure 6.17.

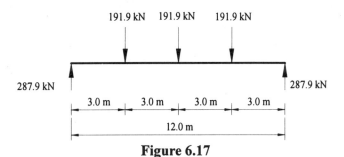

**Figure 6.17**

The estimated maximum deflection is given by      $\delta \approx \dfrac{0.104\ \text{B. Max}\ L^2}{EI}$

In this case

   B.Max =   $(287.9 \times 6) - (191.9 \times 3)$                    =   1151.4 kNm

$$\delta_{max.} \approx \frac{0.104 \times 1151.4 \times 10^3 \times (12000)^2}{205 \times 12.412 \times 10^9} \quad = \quad 6.8 \text{ mm}$$

since   $\delta_{max} < \dfrac{span}{200}$   the deflection is acceptable

### 6.7.10  Intermediate Stiffeners    (Clause 4.4.6)

*Clause 4.4.6.3*

The maximum outstand ($b_s$) of intermediate stiffeners is governed by the requirements of *Clause 4.5.1.2*, i.e.

When the outer edge is not continuously stiffened then      $b_s \leq 19t_s\varepsilon$

and when $13t_s\varepsilon \leq b_s \leq 19t_s\varepsilon$ the design should be based on a core with $b_s = 13t_s\varepsilon$

Assuming 8 mm flats for the stiffeners and $\varepsilon = 1.0$ then   $b_s \leq 19 \times 8 = 152$ mm

The maximum flange width available $= \dfrac{(450-10)}{2} = 220$ mm and is therefore adequate.

*Clause 4.4.6.4*

The minimum second moment of area, $I_s$, about the centre-line of the web should satisfy:

$I_s \geq 0.75d^3$ for  $a \geq \sqrt{2}d$

and

$I_s \geq \dfrac{1.5d^3t^3}{a^2}$  for  $a < \sqrt{2}d$

$a = 750$ mm   $\sqrt{2}\,d = 1414$  $\therefore$  $a < \sqrt{2}d$

$I_s \geq \dfrac{1.5 \times 1000^3 \times 10^3}{750^2} = 2.667 \times 10^6$ mm$^4$

$\dfrac{t_s(2b_s + t)^3}{12} = 2.667 \times 10^6$ mm$^4$

$(2b_s + t) = 158$ mm

outstand $= \dfrac{(158-10)}{2} = 74$ mm

say   75 mm

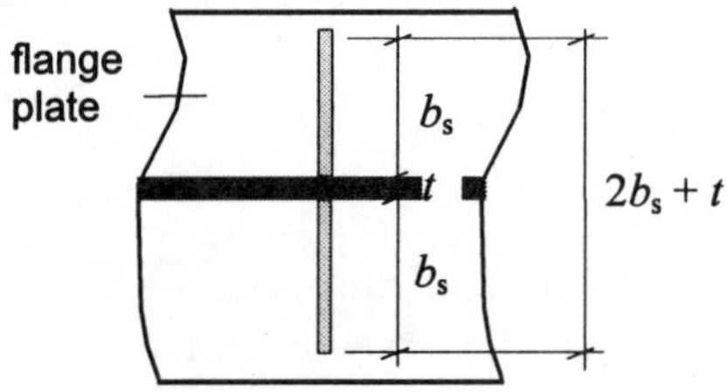

Section through stiffeners
and web

**Figure 6.18**

$13t_s\varepsilon = 13 \times 8 \times 1.0 = 104$ mm $>$ outstand $\therefore$ acceptable

**Two stiffeners 75 mm × 8 mm thick will be adequate.**

*Clause 4.4.6.6*

If the web has been designed using tension-field action it is necessary to carry out a buckling check, otherwise this clause can be ignored.

*Clause 4.4.6.7*

The weld between the web and intermediate stiffeners not subject to external loads should

be capable of resisting a shear force  $\geq \dfrac{t^2}{5b_s}$

$$\therefore \text{ weld capacity } \geq \frac{10^2}{5 \times 75} = 0.27 \text{ kN/mm run on two welds}$$

It is normal practice to use a minimum fillet weld in structural elements of this type of not less than 6 mm, (strength = 0.9 kN/mm).

Adopt four continuous 6 mm fillet welds for each set of intermediate stiffeners. Since the stiffeners are not subject to external loading, they may terminate clear of the tension flange by an amount approximately equal to $4t$, i.e. 40 mm.

The stiffeners should extend to the compression flange but no welding is required at this location.

### 6.7.11  Load Bearing Stiffeners  (Clause 4.5)

*Clause 4.5.4.2*

The minimum area of stiffener which should be in contact with the flange to limit bearing stresses is determined by :

$$\text{contact area } \geq \frac{\left(0.8 \times \text{ the external load or reaction}\right)}{\text{the design strength of the stiffener}}$$

$$A \geq \frac{\left(0.8 \times 1396.2 \times 10^3\right)}{275} = 4061 \text{ mm}^2$$

Assuming stiffeners 12 mm thick
Allowing for a 20 mm fillet for the weld at the web to flange intersection, the contact area $A$ is given by:

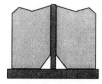

$$A = 2(b_s - 20)12 = (24b_s - 480)$$
$$\text{since } A \qquad\qquad \geq 4061 \text{ mm}^2$$
$$(24b_s - 480) \geq 4061$$
$$_s \qquad\quad \geq 189 \text{ mm}$$

Try stiffeners comprising 2 flats   200 mm $\times$ 12 mm thick.
$$\text{Contact Area } = 2(200 - 20) \times 12 = 4320 \text{ mm}^2$$

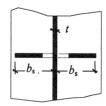

**Figure 6.19**

*Clause 4.5.1.2*

The maximum outstand requirements are the same as those for the intermediate stiffeners.
$$13t_s\varepsilon \quad = 13 \times 12 \times 1.0 = 156 \text{ mm}$$
$$19t_s\varepsilon \quad = 19 \times 12 \times 1.0 = 228 \text{ mm}$$

since $13t_s\varepsilon \leq b_s \leq 19t_s\varepsilon$ the stiffeners should be designed on the basis of a core section equal to $13t_s\varepsilon$, i.e. 156 mm.

*Clause 4.5.1.5*
The buckling resistance of the stiffener is considered to be the compressive resistance of a strut with a cross-section equal to the full or core area of the stiffener together with an effective length of web on each side of the centre-line of the stiffeners, (where available), limited to $20t$. This is shown in Figure 6.20.

The compressive strength $p_c$ is taken from *Table 27(c)* which requires a slenderness ratio

$$\lambda = \frac{\text{effective length}}{\text{radius of gyration}}$$

The radius of gyration of the cross-section should be taken about an axis parallel to the web

$$I = \frac{12 \times 322^3}{12} + \frac{194 \times 10^3}{12} = 33.4 \times 10^6 \text{ mm}^4$$

$$\text{Area} = (322 \times 12) + (194 \times 10) = 5804 \text{ mm}^2$$

$$r_{yy} = \sqrt{\frac{I}{A}} = \sqrt{\frac{33.4 \times 10^6}{5804}} = 78.56 \text{ mm}$$

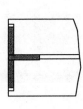

Effective cross-section of stiffener

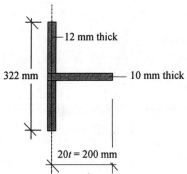

322 mm — 12 mm thick — 10 mm thick — $20t = 200$ mm

**Figure 6.20**

The compressive strength $p_c$ is taken from *Table 27(c)* which requires a slenderness ratio of

$$\lambda = \frac{\text{effective length}}{\text{radius of gyration}}$$

The radius of gyration of the cross-section should be taken about an axis parallel to the web

$$I = \frac{12 \times 322^3}{12} + \frac{194 \times 10^3}{12} = 33.4 \times 10^6 \text{ mm}^4$$

$$\text{Area} = (322 \times 12) + (194 \times 10) = 5804 \text{ mm}^2$$

$$r_{yy} = \sqrt{\frac{I}{A}} = \sqrt{\frac{33.4 \times 10^6}{5804}} = 75.86 \text{ mm}$$

The effective length for load bearing stiffeners where the flange through which the load is being transferred can be considered restrained against lateral movement relative to the other flange should be considered as:

(a) $L_e = 0.7L$
where restraint to the flange against rotation in the plane of the stiffener is provided by other structural members,

(b) $L_e = L$
where no such restraint is provided.

(The effective length for intermediate transverse stiffeners should be assumed to be 0.7L.)
In each of the cases above $L$ is the length of the stiffener.
Assuming case (a) above;

$$L_e = 0.7 \times 1000 = 700 \text{ mm}; \qquad \lambda = \frac{700}{75.86} = 9.23$$

*Clause 4.7.5*
The design strength obtained from *Table 6* should be reduced by 20 N/mm² for welded sections, therefore since $t < 16$ $\qquad p_y = 255$ N/mm².
From *Table 27(c)* when $\lambda = 9.23$ and $p_y = 255$ N/mm² $\qquad p_c = 255$ N/mm²
The buckling resistance of the stiffener $P_x = (255 \times 5804)/10^3 = 1480$ kN

**Since $P_x \geq 1396.2$ kN the stiffeners are adequate with respect to buckling**

*Clause 4.5.5*
The bearing capacity of the stiffeners should be sufficient to sustain the difference between the applied load or reaction and the local capacity of the web, as given in *Clause 4.5.3*. i.e.

Local bearing capacity of the web $= (b_1 + n_2)tp_{yw}$

where:
$b_1$     is the stiff bearing length as defined in *Clause 4.5.1.3* and *Figure 8* of the code
$n_2$     is the length obtained by dispersion of the load through the flange to web connection at a slope of 1:2.5 to the plane of the flange.
$t$     is the web thickness
$p_{yw}$     is the design strength of the web from *Table 6*

Assuming the stiff bearing length to be zero (this is conservative):

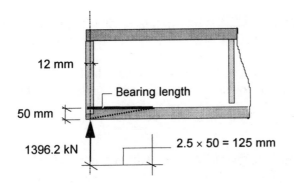
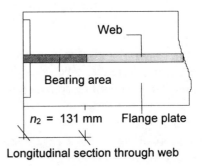

12 mm

50 mm

1396.2 kN

Bearing length

2.5 × 50 = 125 mm

Web

Bearing area

$n_2$ = 131 mm    Flange plate

Longitudinal section through web

**Figure 6.21**

$$P_{\text{local}} = \frac{[(0 + 131)10 \times 275]}{10^3} = 360.2 \text{ kN}$$

The bearing stiffener should be checked for a load  = (reaction − $P_{\text{local}}$)

= 1396.2 − 360.25 = 1036 kN

The bearing capacity of the stiffener                    = $p_{ys}$ × cross-sectional area

The bearing capacity of the stiffener                    $= \dfrac{(275 \times 4320)}{10^3} = 1188 \text{ kN}$

since this is greater than 1036 kN the stiffeners are adequate in bearing.

*Clause 4.5.9*
The connection of the bearing stiffeners to the web should be designed to resist the lesser of:

(i)  the tension capacity of the stiffener, i.e.

$$P_t = A_e p_y = \frac{(5804 \times 275)}{10^3} = 1596 \text{ kN}$$

or
(ii)  the larger of the sum of any forces acting in opposite directions or the sum of any forces acting in the same direction, i.e.

Applied force = End reaction = 1396.2 kN

Case (ii) is the design criterion here,
     Length of the stiffener = 1000 mm

Force to be resisted by the weld / mm = $\dfrac{1396.2}{1000}$ = 1.4 kN/mm

Adopt four continuous 6 mm fillet welds for each set of load bearing stiffeners.

*Clause 4.5.11*
Load bearing stiffeners which are required to resist compression should either be connected by continuous welds or non-slip fasteners or fitted against the loaded flange. In cases where:

        (i)     a load is applied directly over a support;
        (ii)    it forms the end stiffener of a stiffened web      or
        (iii)   it acts as a torsion stiffener  (see *Clause 4.5.8*),

the stiffener should be fitted or connected to both flanges.
It is common practice to specify a full strength weld for the flange/stiffener connection.

### 6.7.10 *Flange to Web Connection*

When a beam/plate-girder bends, in addition to the vertical shear stress at a cross-section there is a complementary horizontal shear stress, i.e.

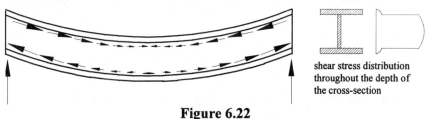

shear stress distribution
throughout the depth of
the cross-section

**Figure 6.22**

The connection between the flange and the web must be designed to transfer this horizontal shear stress '$q$'. The value of '$q$' can be determined from

$$q = \frac{QA\overline{y}}{It} \ \ \text{N/mm}^2$$

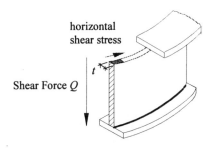

horizontal
shear stress

$t$

Shear Force $Q$

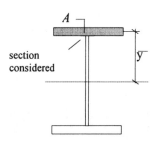

$A$

section
considered

$\overline{y}$

**Figure 6.23**                                       **Figure 6.24**

Shear force / mm length $\quad = \quad q \times t \quad = \dfrac{QA\bar{y}}{I}$ N/mm

where:

$Q$   is the vertical shear force at the position along the span being considered,

$A\bar{y}$   is the first moment of area (about the neutral axis) of the part of the cross-section above the level at which the shear stress is being evaluated,

$I$   is the second moment of area of the cross-section about the neutral axis.

$$A = (450 \times 50) \quad = 22.5 \times 10^3 \text{ mm}^2 \quad \bar{y} = 525 \text{ mm}$$

$$I = \frac{450 \times 1100^3}{12} - \frac{440 \times 1000^3}{12} = 12.412 \times 10^9 \text{ mm}^4$$

$Q$ shear at end $\quad = 1396.2$ kN

Shear force / mm $\quad = \dfrac{1396.2 \times 22.5 \times 10^3 \times 525}{12.412 \times 10^9} = 1.33$ kN/m

The strength of two 6 mm fillet welds $= (2 \times 0.903) \quad = 1.806$ kN/mm.

**Adopt a continuous 6 mm fillet weld on each side of the web.**

**Note:** An intermittent weld could have been used here but this tends to create potential corrosion pockets and is best avoided.

## 6.8 Example 6.2 Plate girder with intermittent restraint to flange

A welded plate girder is simply-supported over a span of 15.0 m and supports two columns at the third-span points in addition to a uniformly distributed load along its length. Adequate lateral restraint is provided to the compression flange at the supports and column locations. Using the data provided design a suitable plate girder.

**Design data:**
Column loads
   service dead load     $= 140$ kN
   service imposed load    $= 400$ kN
Uniformly distributed load
   service dead load     $= 8.0$ kN/m
   service imposed load    $= 25.0$ kN/m
Assumed self-weight of girder   $= 3.0$ kN/m

Solution to Example 6.2   (see Section 6.9).

## 6.9 Solution to Example 6.2

| Contract : Connections  Job Ref. No. : Example 6.2<br>Part of Structure :   Plate  Girders<br>Calc. Sheet No. :  1  of 11 | Calcs. by : W.McK.<br>Checked by :<br>Date : |
| --- | --- |

| References | Calculations | Output |
| --- | --- | --- |
| *BS 5950:Pt 1*<br>*BS 6399:Pt 1*<br><br><br><br><br><br><br><br><br>*Table 2*   $\gamma_f$ | *Structural  use of steelwork in building*<br>*Loading for buildings, Code of practice for imposed loads*<br><br>Unfactored U.D. Dead Load $\qquad = \quad$ 8.0 kN<br>Unfactored U.D. Imposed Load $\qquad = \quad$ 25.0 kN<br>Assumed self-weight of girder $\qquad = \quad$ 3.0 kN<br>Unfactored Point Load (dead) $\qquad = \quad$ 140 kN<br>Unfactored Point Load (imposed) $\quad = \quad$ 400 kN<br>Design Loads<br>U.D.L.. $\quad = [1.4 \times (8+3)] + [1.6 \times 25] \quad = \quad$ 55.4 kN/m<br>Point Loads $= [1.4 \times 140] + [1.6 \times 400] \quad = \quad$ 836 kN | |

836 kN     836 kN     55.4 kN/m

1251.5 kN   5.0 m   5.0 m   5.0 m   1251.5 kN

15.0 m

For analysis this can be considered as the  superposition
of  two separate load cases as shown below :

55.4 kN/m

415.5 kN   5.0 m   5.0 m   5.0 m   415.5 kN

15.0 m

836 kN   836 kN

836 kN   5.0 m   5.0 m   5.0 m   836 kN

15.0 m

Support reactions $\qquad = \quad (415.5 + 836) \qquad = \quad$ 1251.5 kN
Design shear force $\qquad\qquad\qquad\qquad = \quad$ 1251.5 kN

| Contract : Connections  Job Ref. No. : Example 6.2<br>Part of Structure :   Plate Girders<br>Calc. Sheet No. : 2  of  11 | Calcs. by : W.McK.<br>Checked by :<br>Date : |
|---|---|

| References | Calculations | Output |
|---|---|---|

Design bending moment $= M_x$

$$M_x = \left(\frac{55.4 \times 15^2}{8}\right) + (836 \times 5) \quad = \quad 5738.125 \text{ kNm}$$

Shear Force Diagram

5565 kN   5738.125 kN   5565 kN

Bending Moment Diagram

Initial sizing        Assume a span/depth ratio of 14

depth of girder $d \;\approx\; \dfrac{15000}{14} = \; 1071$ mm

say  1200 mm

Assuming    $16 \le T \le 40$  mm

*Table 6*       $p_y = 265$ N/mm$^2$

Flange force   $\approx \; \dfrac{5738 \times 10^3}{1200} = \; 4781$ kN

*Clause 4.3.7.3*   Since the flange is not fully restrained a value less than 265 N/mm$^2$ should be used when estimating the required flange area. The moment capacity will be based on:

$$M_b = p_b \times S_{xx}$$

Assume    $p_b = 260$ N/mm$^2$
Area of flange plate req'd

$$A_f \;\approx\; \dfrac{4781 \times 10^3}{260} = \; 18.38 \times 10^3 \text{ mm}^2$$

| Contract : Connections  Job Ref. No. : Example 6.2 | Calcs. by : W.McK. |
|---|---|
| Part of Structure :   Plate Girders | Checked by : |
| Calc. Sheet No. : 3 of 11 | Date : |

| References | Calculations | Output |
|---|---|---|
| | Try flange plates 460 mm wide x 40 mm thick. | |

Try flange plates 460 mm wide x 40 mm thick.
$A_{provided} = 460 \times 40 = 18.4 \times 10^3$ mm$^2$
Assume a 10 mm thick web
Trial section :

Section properties of proposed girder:
Area $= 2( 460 \times 40 ) + ( 10 \times 1200 ) = 48.8 \times 10^3$ mm$^2$

$$I_{xx} = \left[ \frac{460 \times 1280^3}{12} \right] - \left[ \frac{450 \times 1200^3}{12} \right]$$

$$= 15.59 \times 10^9 \text{ mm}^4$$

$$I_{yy} = 2\left[ \frac{40 \times 460^3}{12} \right] + \left[ \frac{1200 \times 10^3}{12} \right]$$

$$= 649 \times 10^6 \text{ mm}^4$$

$$r_{yy} = \sqrt{\frac{I_{yy}}{A}} = \sqrt{\frac{649 \times 10^6}{48.8 \times 10^3}} = 115.32 \text{ mm}$$

$$S_{xx} = ( 460 \times 40 ) \times ( 1240 ) = 22.82 \times 10^6 \text{ mm}^3$$

$$\text{Self-weight} = \left(48.8 \times 10^3\right) \times (1000) \times \left(\frac{7.85}{10^8}\right) = 3.8 \text{ kN/m}$$

(this is more than the assumed value, but is o.k.)

*Table 7*   Section Classification
Flanges :

*Table 6*   $T = 40$ mm     $p_y = 265$ N/mm$^2$     $\varepsilon = \left(\dfrac{275}{265}\right)^{\frac{1}{2}} = 1.02$

| Contract : Connections  Job Ref. No. : Example 6.2<br>Part of Structure :   Plate Girders<br>Calc. Sheet No. :  4  of  11 | Calcs. by : W.McK.<br>Checked by :<br>Date : |
| --- | --- |

| References | Calculations | Output |
| --- | --- | --- |
| | $b = \left(\dfrac{460 - 10}{2}\right) = 225$ mm | |
| | $\dfrac{b}{T} = \dfrac{225}{40} = 5.63 \leq 7.5\varepsilon$ | |
| | Web : | |
| | $t = 10$ mm    $p_y = 275$ N/mm$^2$       $\varepsilon = \left(\dfrac{275}{275}\right)^{\frac{1}{2}} = 1.0$ | |
| | $\dfrac{d}{t} = \dfrac{1200}{10} = 120$    $98\varepsilon \leq \dfrac{d}{t} \leq 120\varepsilon$ | |
| Clause 3.6.2 | $\dfrac{d}{t} > 63\varepsilon$  web is thin, use *Clause 4.4.4.2* to determine<br>moment capacity | |
| Clause 4.4.2.2 | b)  with transverse stiffeners only :<br><br>where stiffener spacing $a > d$  then   $t \geq \dfrac{d}{250}$<br><br>where stiffener spacing $a \leq d$  then   $t \geq \dfrac{d}{250}\left(\dfrac{a}{d}\right)^{\frac{1}{2}}$<br><br>Assuming the more critical case with stiffeners,<br>then   $t \geq \dfrac{1200}{250} = 4.8$ mm | |
| Clause 4.4.2.3 | Assuming the more critical case with stiffeners<br>then   $t \geq \dfrac{1200}{250}\left(\dfrac{265}{345}\right) = 3.7$ mm | Web is adequate<br>with respect to<br>serviceability |
| Clause 4.3.7.3<br>Table 11 | $M_b = S_{xx}.p_b$<br>$p_b$ is dependent on $p_y$ and  equivalent  slenderness $\lambda_{LT}$<br>$\qquad\qquad\quad \lambda_{LT} = nuv\lambda$<br><br>$\lambda = \dfrac{L_E}{r_{YY}} = \dfrac{1500}{115.32} = 43.36$ | $\lambda = 43.36$ |
| Appendix B<br>Clause B.2.5 | For sections which are symmetrical about both axes,<br>either set of equations can be used to determine | |

| References | Calculations | Output |
|---|---|---|
| | **'U' and 'x'** $\quad U = \left(\dfrac{4S_x^2\gamma}{A^2 h_s^2}\right)^{0.25}$ | |
| | where: | |
| | $S_{xx} = 22.82 \times 10^6 \text{ mm}^3$ | |
| | $\gamma = \left(1 - \dfrac{649 \times 10^6}{15.59 \times 10^9}\right) = 0.958$ | |
| | $A = 48.8 \times 10^3 \text{ mm}^2$ | |
| | $h_s = 1200 + 40 = 1240 \text{ mm}$ | |
| | $U = \left(\dfrac{4 \times \left(22.82 \times 10^6\right)^2 \times 0.985}{\left(48.8 \times 10^3\right)^2 \times 1240^2}\right)^{0.25} = 0.859$ | $U = 0.859$ |
| | $X = 0.566 h_s \left(A/J\right)^{\frac{1}{2}}$ | |
| | where : | |
| | $J = \dfrac{1}{3}(2T^3 B + t^3 d)$ | |
| | $J = \dfrac{1}{3}\left[\left(2 \times 40^3 \times 460\right) + \left(10^3 \times 1200\right)\right]$ | |
| | $\quad = 20.03 \times 10^6 \text{ mm}^4$ | |
| | $X = 0.566 \times 1240 \times \left(\dfrac{48.8 \times 10^3}{20.03 \times 10^6}\right)^{\frac{1}{2}} = 34.64$ | $X = 34.64$ |
| *Table 14* | $\dfrac{\lambda}{x} = \dfrac{43.36}{34.64} = 1.25; \quad N = 0.5$ | $v = 0.98$ |
| *Clause 4.3.7.6* | Assuming that the plate girder is not subject to destabilising loads | |
| *Table 13* | $m = 1.0$ | |
| *Table 17* | $\beta$ is +ve. and $\gamma$ is +ve | |
| | $M_o = \dfrac{55.4 \times 5^2}{8} = 173.125$ | |
| *Table 16* | $\gamma = \dfrac{M}{M_o} = \dfrac{5565}{173.125} = 32.14$ | $n = 0.98$ |
| | $\lambda_{LT} = nuv\lambda = 0.98 \times 0.859 \times 0.98 \times 43.36 = 35.77$ | $\lambda = 35.77$ |

Contract : Connections Job Ref. No. : Example 6.2
Part of Structure : Plate Girders
Calc. Sheet No. : 5 of 11

Calcs. by : W.McK.
Checked by :
Date :

| Contract : Connections  Job Ref. No. : Example 6.2<br>Part of Structure :   Plate Girders<br>Calc. Sheet No. :  6  of  11 | Calcs. by : W.McK.<br>Checked by :<br>Date : |
| --- | --- |

| References | Calculations | Output |
| --- | --- | --- |
| *Table 16* | For $p_y = 265$ N/mm$^2$   and   $\lambda_{LT} = 35.77$   $p_b = 263$ N/mm$^2$<br>$M_b = S_{xx} \cdot p_b$<br>$M_b = (263 \times 22.82 \times 10^6)/10^6 = \quad$ 6001.7 kNm.<br>$\qquad\qquad\qquad M_b > M_x$ | Section is adequate with respect to bending |
| *Clause 4.4.5.3* | Shear Capacity :<br>Outer third of span<br>$V_{cr} = q_{cr} dt$   where $V_{cr} = 1251.5$ kN<br><br>Min. $q_{cr}$ required   $\geq \dfrac{V_{cr}}{dt} = \dfrac{1251.5 \times 10^3}{1200 \times 10} = 104.3$ N/mm$^2$ | |
| *Table 21(b)* | $p_y = 275$ N/mm$^2$ , $d/t = 120$,<br>$\therefore$  $a/d < $ 1.2 ( $q_{cr} = 106$ N/mm$^2$ )<br>Stiffener spacing  $a \leq 1.2 \times 1200 = 1440$ mm<br>A stiffener spacing of 1250 mm is suitable for the outer third. | Adopt stiffeners @ 1250 mm centres for the outer third |
| | Middle third of span  $V_{cr} = 138.5$ kN<br><br>Min. $q_{cr}$ required  $\geq \dfrac{138.5 \times 10^3}{1200 \times 10} = 11.54$ N/mm$^2$ | |
| *Table 21(b)* | $a/d = \infty$     $q_{cr} = 69$ N/mm$^2$ | |
| | Intermediate stiffeners are not required in the middle third<br>At this point the web serviceability requirements should be re-checked and the web between the stiffeners should be checked for edge-loading. | |
| *Clause 4.4.2.2* | $t \geq \dfrac{d}{250} = \dfrac{1200}{250} = 4.8$ mm | |
| *Clause 4.4.2.3* | $t \geq \dfrac{d}{250}\left(\dfrac{p_{yf}}{345}\right) = \dfrac{1200}{250}\left(\dfrac{265}{345}\right) = 3.7$ mm | |
| | Assuming no rotational restraint to top flange | |
| *Clause 4.5.2.2* | $f_{ed} \leq p_{ed}\left[1.0 + \dfrac{2}{(a/d)^2}\right]\dfrac{E}{(d/t)^2}$<br><br>$f_{ed} = \dfrac{55.4}{10} = 5.5$ N/mm$^2$ | |

| References | Calculations | Output |
|---|---|---|
| | **Contract : Connections  Job Ref. No. : Example 6.2**<br>**Part of Structure :Plate Girders**<br>**Calc. Sheet No. :  7  of  11** | **Calcs. by : W.McK.**<br>**Checked by :**<br>**Date :** |

| References | Calculations | Output |
|---|---|---|
| | Assuming no stiffeners in the middle third,<br>$a/d = \infty$<br><br>$P_{ed} = \dfrac{E}{(d/t)^2} = \dfrac{205 \times 10^3}{120^2} = 14.24 \text{ N/mm}^2$<br><br>$p_{ed} > f_{ed}$ | No stiffeners are<br>required in the<br>middle third. |
| | <br>5000 mm<br><br>4 @1250 crs. = 5000mm       4 @1250 crs. = 5000mm | |
| *Clause 4.2.5* | Deflection<br><br>$\delta_{max} \leq \dfrac{span}{360} = \dfrac{15000}{360} = 41.7 \text{ mm}$<br><br>$\delta_{udl} = \dfrac{5Wl^3}{384EI}$ and $\delta_{point\ loads} = \dfrac{23Pl^3}{648EI}$<br><br>Deflection due to unfactored imposed loads<br>UDL   $W = 25$ kN/m        Point loads $P = 400$ kN<br><br>$\delta_{act} \approx \dfrac{l^3}{EI}\left[\dfrac{5W}{384} + \dfrac{23P}{648}\right]$<br><br>$\delta_{act} \approx \dfrac{15000^3}{205 \times 15.59 \times 10^9}\left[\dfrac{5 \times 25 \times 15}{384} + \dfrac{23 \times 400}{648}\right]$<br><br>$\delta_{act} \approx 20.2 \text{ mm} < 41.7 \text{ mm}$<br><br>or<br><br>Using the equivalent UDL technique,<br>B.Max. due to unfactored loads        $= 2703$ kNm.<br><br>$\delta_{max} \approx \dfrac{0.104 \times 2703 \times 10^3 \times 15000^2}{205 \times 15.59 \times 10^9} = 19.8 \text{ mm}$ | Deflection is<br>acceptable |
| *Clause 4.5.1.2*<br>*Table 6* | Intermediate stiffeners<br>Assume 8 mm thick flats   $p_y = 275 \text{ N/mm}^2$   $\varepsilon = 1.0$ | |

| Contract : Connections  Job Ref. No. : Example 6.2 | Calcs. by : W.McK. |
|---|---|
| Part of Structure :  Plate Girders | Checked by : |
| Calc. Sheet No. :  8  of  11 | Date : |

| References | Calculations | Output |
|---|---|---|
| | $\therefore$  Outstand $b_s$  $\leq$  $19 \times 8$ $= 152$ mm | |
| | Maximum flange width available $= \dfrac{(460 - 10)}{2} = 225$ mm | Adequate |
| Clause 4.4.6.4 | $a = 1250$ mm,  $\sqrt{2}d$ $=$ $\sqrt{2} \times 1200$ $=$ $1697$ mm | |
| | $a$  $<$  $\sqrt{2}d$ | |
| | $I_s$  $\geq$  $\dfrac{1.5 d^3 t^3}{a^2} = \dfrac{1.5 \times 1200^3 \times 10^3}{1250^2} = 1.659 \times 10^6$ mm$^4$ | |
| | $I_s$  $=$  $\dfrac{t_s (2b_s + t)^3}{12}$ $=$ $1.659 \times 10^6$ | |
| | $(2b_s + t) = 135.5$  $\therefore$ $b_s = 62.8$ mm  say  65 mm | Adopt 2/stiffeners |
| | $13 t_s \varepsilon$ $= 13 \times 8 \times 1.0$ $=$ $1.4 >$  outstand | 65 mm $\times$ 8 mm thick |
| Clause 4.4.6.7 | Weld capacity $\geq$ $\dfrac{t^2}{5 b_s}$ $=$ $\dfrac{10^2}{5 \times 65}$ | Adopt 4 continuous |
| | $= 0.31$ kN/mm on two welds | 6 mm fillet welds |
| | Terminate stiffeners at 4t clear of tension flange | |
| | i.e.  $4 \times 10 = 40$ mm | |
| | Extend stiffeners to compression flange, no weld required. | |
| | Load bearing stiffeners under point loads | |
| Clause 4.5.4.2 | Contact area $A$  $>$  $\dfrac{0.8 \times 836 \times 10^3}{275}$ $=$ $2432$ mm$^2$ | |
| | Assuming stiffeners 10 mm thick, and allowing 20 mm fillet for web/flange weld | |
| | $A$ $=$ $2(b_s - 20) \times 10 = 2432$ | |
| | $b_s$ $=$ $142$ mm | |
| Clause 4.5.1.2 | Try stiffeners comprising 2 flats 145 $\times$ 10 mm thick | |
| | $13 t_s \varepsilon$ $=$ $(13 \times 10 \times 1.0) = 130$ mm | |
| | $19 t_s \varepsilon$ $=$ $(19 \times 10 \times 1.0) = 190$ mm | |
| | $130 < b_s < 190$ | |
| | $\therefore$ Use core section equal to 130 mm | |

| Contract : Connections  Job Ref. No. : Example 6.2<br>Part of Structure :   Plate girders<br>Calc. Sheet No. :  9  of  11 | Calcs. by : W.McK.<br>Checked by :<br>Date : |
|---|---|

| References | Calculations | Output |
|---|---|---|
| *Clause 4.5.1.5* |

10 mm thick

10 mm thick

20t = 200 mm    20t = 200 mm

$I = \dfrac{10 \times 270^3}{12} + \dfrac{400 \times 10^3}{12} = 16.44 \times 10^6 \text{ mm}^4$

Area $= (270 \times 10) + (400 \times 10) = 6700 \text{ mm}^2$

$r = \sqrt{\dfrac{16.44 \times 10^6}{6700}} = 49.5 \text{ mm}$

$L_E = 0.7 \times 1200 = 840 \text{ mm}$

$\lambda = \dfrac{840}{49.5} = 16.91$ | |
| *Clause 4.7.5*<br>*Table 27(c)* | $p_y = 255 \text{ N/mm}^2$  (*Table 6* value less 20 N/mm²)<br>$p_c = 253 \text{ N/mm}^2$ | |
|  | Buckling resistance $= P_x = (253 \times 6700)/10^3$<br>$\qquad\qquad\qquad\quad P_x = 1695 \text{ kN} \gg 836 \text{ kN}$ | Adequate in buckling |
| *Clause 4.5.5* | Bearing capacity $\geq$ (Applied load $- P_{local}$)<br>$\qquad\qquad\qquad = (836 - 275) = 561 \text{ kN}$<br>$P_{local} = (b_1 + n_2)tp_{yw}$<br>Assume $b_1 = 0$   $n_2 = (2.5 \times 40) = 100 \text{ mm}$<br>$P_{local} = (100 \times 10 \times 275)/10^3 = 275 \text{ kN}$<br><br>Bearing capacity $= P_{crip} = (275 \times 6700)/10^3$<br>$\qquad\qquad\qquad\qquad P_{crip} = 1843 \text{ kN} \gg 561 \text{ kN}$ | Adequate in bearing |

| References | Calculations | Output |
|---|---|---|

**Contract : Connections  Job Ref. No. : Example 6.2**
**Part of Structure :  Plate Girders**
**Calc. Sheet No. :  10  of  11**

**Calcs. by : W.McK.**
**Checked by :**
**Date :**

---

*Clause 4.5.9*

Welded connection

Tension capacity    $P_t = A_e p_y$    $= (6700 \times 275)/10^3$
                                        $= 1843$ kN

Applied force                         $= 836$ kN
Design weld for 836 kN

Length of stiffener  = 1200 mm

Strength of weld required $= \dfrac{836}{1200} = 0.7$ kN/mm

**Output:** Adopt 4 continuous

6 mm fillet welds

End Post

*Clause 4.5.4.2*

Contact area  $A > \dfrac{0.8 \times 1251.5 \times 10^3}{275} = 3641$ mm$^2$

Assuming stiffeners 12 mm thick, and allowing 20 mm fillet for web/flange weld
$A = 2(b_s - 20) \times 10 = 3641$ mm$^2$
$b_s = 172$ mm
Try stiffeners comprising 2 flats 180 × 12 mm thick

*Clause 4.5.1.2*

$13 t_s \varepsilon = (13 \times 12 \times 1.0) = 156$ mm
$19 t_s \varepsilon = (19 \times 12 \times 1.0) = 228$ mm
              $130 < b_s < 190$
∴ Use core section equal to 156 mm

*Clause 4.5.1.5*

12 mm thick
322 mm
10 mm thick
20t = 200 mm

$I = \dfrac{10 \times 322^3}{12} + \dfrac{206 \times 10^3}{12} = 33.41 \times 10^6$ mm$^4$

Area $= (322 \times 10) + (246 \times 10) = 6324$ mm$^2$

$r = \sqrt{\dfrac{33.41 \times 10^6}{6324}} = 72.7$ mm

| References | Calculations | Output |
|---|---|---|
| | $L_E$ = $(0.7 \times 1200)$ = 840 mm | |
| | $\lambda$ = $\dfrac{840}{72.7}$ = 11.6 | |
| *Clause 4.7.5* | $p_y$ = 255 N/mm² (*Table 6* value less 20 N/mm²⁾) | |
| *Table 27(c)* | $p_c$ = 255 N/mm² | |
| | Buckling resistance = $P_x$ = $(255 \times 6324)/10^3$ | |
| | $P_x$ = 1613 kN > 1251.5 kN | Adequate in buckling |
| | Bearing capacity ≥ (Applied load − $P_{local}$) | |
| | $P_{local}$ = $(b_1 + n_2)tp_{yw}$ = 275 kN as before | |
| | (Applied load − $P_{local}$)= (1251.5 − 275) = 976.5 kN | |
| *Clause 4.5.3* | Bearing capacity $P_{crip}$ = $(275 \times 6324)/10^3$ = 1739 kN | |
| | $P_{crip}$ >> 976.5 kN | Adequate in bearing |
| *Clause 4.5.9* | Welded connection | |
| | Tension capacity $P_t$ = $A_e p_y$ = $(6324 \times 275)/10^3$ | |
| | = 1739 kN | |
| | Applied force = 1251.5 kN | |
| | Design weld for 1251.5 kN | |
| | Length of stiffener = 1200 mm | Adopt 4 continuous |
| | Strength of weld required = $\dfrac{1251.5}{1200}$ = 1.04 kN/mm | 6 mm fillet welds |
| | Flange to web connection | |
| | $q$ = $\dfrac{QA\overline{y}}{I}$ N/mm² | |
| | $Q$ = 1251.5 kN | |
| | $A\overline{y}$ = $(460 \times 40 \times 620)$ = $11.408 \times 10^6$ mm³ | |
| | $I_{xx}$ = $15.59 \times 10^9$ mm⁴ | Adopt 2 continuous |
| | $q$ = $\dfrac{1251.5 \times 11.408 \times 10^6}{15.59 \times 10^9}$ = 0.915 kN/mm | 6 mm fillet welds |

**Contract : Connections  Job Ref. No. : Example 6.2**
**Part of Structure :   Plate Girders**
**Calc. Sheet No. :  11  of  11**

**Calcs. by : W.McK.**
**Checked by :**
**Date :**

# 7. Concise Eurocode 3 C-EC3

## 7.1 Introduction

This chapter provides an introduction to the contents of the Concise Eurocode 3 and illustrates the design of simple structural elements. A more comprehensive treatment can be found in *Introduction to Concise Eurocode 3 (C-EC3) – with Worked Examples* published by The Steel Construction Institute.

The European Standards Organisation, CEN, is the umbrella organisation under which a set of common structural design standards (e.g. EC1, EC2, EC3, etc.) have been developed. The Structural Eurocodes are the result of attempts to eliminate barriers to trade throughout the European Union. Separate codes exist for each structural material, including EC 3 for steel. The basis of design and loading considerations is included in EC1.

Each country publishes its own European Standards (EN); for example, in the UK the British Standards Institution (BSI) issues documents, which are based on the Eurocodes developed under CEN, with the designation BS EN.

Currently, the Structural Eurocodes are issued as Pre-standards (ENV) which can be used as an alternative to existing national rules. In the UK the BSI has used the designation DD ENV; the pre-standards are equivalent to the traditional 'Draft for development' Documents.

In the UK the Eurocode for steelwork design is known as 'DD ENV 1993-1-1 Eurocode 3: Design of steel structures: Part 1.1 General rules for buildings' (together with the United Kingdom National Application Document).

Eurocode 3 adopts the 'Limit State Design' philosophy as currently used in UK national standards.

### National Application Document    (NAD)

Each country which issues a European Standard also issues a NAD for use with the EN. The purpose of the NAD is to provide information to designers relating to product standards for materials, partial safety factors and any additional rules and/or supplementary information specific to design within that country.

### Concise EC3    (C-EC3)

In the UK The Steel Construction Institute has published *C-EC3 – Concise Eurocode 3 for the Design of Steel Buildings in the United Kingdom*. This document is an abridged version of DD ENV 1993-1-1 and covers only the sections and Clauses which are necessary for the design of types of steel structures most commonly used in the UK. This can be used as a stand-alone design standard for most building structures. Unlike EC 3, many of the design provisions in this document are presented in Tabular or Chart format which will be familiar to UK designers. A summary of the abbreviations is given in Table 7.1.

**Table 7.1**

| Abbreviation | Meaning |
|:---:|:---|
| **CEN** | European Standards Organisation |
| **EC** | Eurocode produced by CEN |
| **EN** | European Standard based on Eurocode and issued by member countries |
| **ENV** | Pre-standard of Eurocode issued by member countries |
| **DD ENV** | U.K version of Pre-standard1 (BSI) |
| **NAD** | National Application Document issued by member countries (BSI) |
| **C-EC 3** | Concise Eurocode (abridged version of DD ENV 1993-1-1) published by the SCI for UK designers |

## 7.2  Terminology, Symbols and Conventions

The terminology, symbols and conventions used in EC 3 differ from those used by BS 5950:Part 1. The code indicates **'Principles'** which are general statements and definitions which must be satisfied and **'Rules'** which are design procedures which comply with the principles. The rules can be substituted by alternative procedures provided that they can be shown to be in accordance with the principles; C-EC3 does not make this distinction.

There are two types of Annexe in EC3: normative and informative. Normative Annexes have the same status as the main body of the text, while Informative Annexes provide additional information. The Annexes generally contain more detailed material or material which is used less frequently.

### 7.2.1  Decimal Point

Standard ISO practice has been adopted in representing a decimal point by a comma, i.e. $5,3 \equiv 5.3$.

### 7.2.2  Actions

The term 'actions' is used when referring to:

- permanent loads   (*G*)   e.g.   dead loads such as self-weight
- variable loads   (*Q*)   e.g.   imposed, wind and snow loads
- accidental loads   (*A*)   e.g.   explosions, fire

### 7.2.3  Resistance

The term 'resistance' is used when referring to member capacity:

- shear resistance        $(V_{pl \cdot Rd})$
- moment resistance       $(M_{c \cdot Rd})$
- buckling resistance      $(R_{b \cdot Rd})$

### 7.2.4  Subscripts

Multiple subscripts are used to denote variables and are defined in *Appendix B* of C-EC3. The level of sophistication is above that used in BS 5950:Part 1; for example, $V_{pl \cdot Rd}$ is the design (*d*), plastic (*pl*), shear(*V*) resistance (*R*) of a section.

### 7.2.5  Design Values

The term 'design' is used for factored loading and member resistance.

Design loading $(F_d)$ = characteristic (*k*) value × partial safety factor $(\gamma_F)$

Design resistance $(R_d)$ $= \dfrac{\text{characteristic } (k) \text{ value}}{\text{partial safety factor } (\gamma_M)}$

**Note:** Design loads apply to both serviceability and ultimate limit states.
The application of design loads to a structure results in design effects such as internal forces and moments denoted by $S_d$. Structures must be designed such that

$$S_d \leq R_d$$

### 7.2.6  Partial Safety Factors

The Eurocode provides indicative values for various safety factors and are shown in the text as 'boxed values', i.e. $\boxed{1,35}$. Each country defines 'boxed values' within the NAD document to reflect the levels of safety required by the appropriate authority of the national government; in case of the UK, the BSI.

A comparison between the values of partial safety factor for loads which are adopted in C-EC 3 and BS 5950:Part 1 is given in Table 7.2.

**Table 7.2**

| Load Type | C-EC 3 | BS 5950:Part 1 |
|---|---|---|
| Permanent  $(G)$ | $\gamma_G = 1.35G$ | $\gamma_F = 1.4G$ |
| Variable     $(Q)$ | $\gamma_Q = 1.5Q$ | $\gamma_F = 1.6Q$ |
| (Wind) | includes wind load | $\gamma_F = 1.4W$ |
| Permanent + Variable (+ Wind) | $1{,}35G + 1{,}35Q$ $+ (0{,}9 \times 1{,}35)Q_{\text{wind}}$ | $1.2G + 1.2Q$ $+ 1.2W$ |

The UK NAD includes a rule (*Clause 4(b)*) which states that the wind loading should be taken as 90% of the value obtained from CP 3: Chapter V: Part 2: 1972. This compensates for the higher value of $\gamma_Q$ used in C-EC 3. **Note:** CP 3: Chapter V: Part 2: 1972 has been superseded by BS 6399: Part 2 1995.

When considering combinations of loads consisting of more than one variable load, EC3 adopts a method utilizing a combination factor $\psi$. The values of $\psi$ for different types of variable load are given in EC1, but are currently included in the UK NAD. For most building structures in normal design situations, a simplified approach is also given in EC3. This is reflected in the modified partial load factor of 1,35 for the combination of more than one independent variable load. In addition to the above, imposed floor loads and imposed roof loads are considered to be independent types of variable load. This simplified approach has been adopted in C-EC3.

The partial safety factors for resistance $\gamma_M$ are given in *Clause 5.1* of C-EC3 and, unlike BS 5950: Part 1 where $\gamma_M = 1.0$, there are three values:

$\gamma_{mo}$ = 1,05  applies when plastic yielding occurs at failure

$\gamma_{m1}$ = 1,05  applies when either overall or local buckling occurs at failure

$\gamma_{m2}$ = 1,2   applies when failure occurs on the net section of bolt holes

Separate values apply for connection design, details of which are given in *Chapter 6* of C-EC3.

### 7.2.7 Symbols

The symbols used to denote section properties in C-EC3 are different from those used in BS 5950: Part 1 and the current structural section property tables published by SCI. A number of 'Properties of Structural Steel Sections' tables are included in Part IV of *Introduction to Concise Eurocode 3 (C-EC3) – with Worked Examples* published by SCI (publication number P115). Tables 7.3 and 7.4 give comparisons of the symbols used in C-EC3 and BS 5950:Part 1.

**Table 7.3**

| Code | Depth of section | Width of section | Thickness | | Root radius |
|------|------------------|------------------|-----------|---|-------------|
| | | | Web | Flange | |
| C-EC3 | $h$ | $b$ | $t_w$ | $t_f$ | $r$ |
| BS 5950:Part 1 | $D$ | $B$ | $t$ | $T$ | $r$ |
| Code | Depth between fillets | Shear area | Ratio for local buckling | | Area |
| | | | | | |
| C-EC3 | $d$ | $A_v$ | $c/t_f$ | $d/t_w$ | $A$ |
| BS 5950:Part 1 | $d$ | $A_v$ | $b/T$ | $d/t$ | $A$ |

**Table 7.4**

| Code | Second moment of area | | Radius of gyration | | Lateral torsional buckling constants | | |
|------|------|------|------|------|------|------|------|
| | major | minor | major | minor | | | |
| C-EC3 | $I_{yy}$ | $I_{zz}$ | $i_{yy}$ | $i_{zz}$ | $i_{LT}$ | $a_{LT}$ | $a_{LT}/i_{LT}$ |
| BS 5950:Part 1 | $I_{xx}$ | $I_{yy}$ | $i_{xx}$ | $i_{yy}$ | see Note * | | |
| Code | Elastic modulus | | Plastic modulus | | Warping constant | Torsion constant | |
| | major | minor | major | minor | | | |
| C-EC3 | $W_{el \cdot y}$ | $W_{el \cdot z}$ | $W_{pl \cdot y}$ | $W_{pl \cdot y}$ | $I_w$ | $I_t$ | |
| BS 5950:Part 1 | $Z_{xx}$ | $Z_{yy}$ | $S_{xx}$ | $S_{yy}$ | $H$ | $J$ | |

**Note:** Two new parameters $i_{LT}$ and $a_{LT}$ are defined in *Clauses (16)* and *(17)* of C-EC3. The parameters combine the term $r_y$ with the buckling parameter '$U$' and the torsional index '$x$' used in BS 5950: Part 1. The value of $a_{LT}/i_{LT}$ is approximately equal to $0,8h/t_f$.

### 7.2.8 Conventions

The difference in conventions most likely to cause confusion for UK engineers is the change in the symbols used to designate the major and minor axes of a cross-section. Traditionally, in the UK the **y-y axis** has represented the minor axis; in C-EC3 this represents the **major axis,** the minor axis being represented by the z-z axis. The **x-x axis** defines the **longitudinal axis**. All three axes are shown in Figure 7.1.

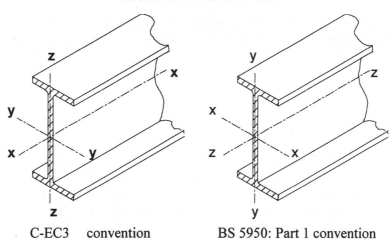

C-EC3   convention        BS 5950: Part 1 convention

**Figure 7.1**

### 7.2.9 *Materials*

The grades of steel given in *Table 3.1* of C-EC3 are designated as *Fe 430* or *S275*, both of which represent steel equivalent to *grade 43* and *Fe 510* or *S355* which represent steel equivalent to *grade 50*. The two designations '*Fe*' and '*S*' represent the designations given to types of steel in BS EN 10025:1990 and BS EN10025:1993 respectively; the 1993 version supersedes the earlier 1990 version. Both designations are given in C-EC3.

The numbers following *Fe* and *S* are the nominal ultimate tensile strength $f_u$ and yield strength $f_y$ respectively, i.e.

$$\text{for} \quad \textit{Fe 430 or S275} \text{ steel} \quad \begin{aligned} f_u &= 430 \text{ N/mm}^2 \\ f_y &= 275 \text{ N/mm}^2 \end{aligned}$$

These values must be adopted as characteristic values for design. Generally, $f_y$ will be used in calculations; however, there are circumstances, such as determining the tensile resistance of the net area at bolt holes, when $f_u$ will be used.

Two categories of material thickness are given in *Table 3.1* of C-EC3:

$$t \;\leq\; 40 \text{ mm} \quad \text{and} \quad 40 \text{ mm} \;<\; t \;\leq\; 100 \text{ mm}$$

when determining the characteristic material strength, four are given in Table 6 of BS 5950: Part 1.

Although not stated in EC3, for rolled I and H sections '$t$' is normally assumed to be the flange thickness. When considering the web resistance, the value of $f_y$ corresponding to the web thickness can be used as in BS 5950: Part 1.

In *Clause 3.1.4* of C-EC3 the value of the Modulus of Elasticity ($E$) is given as 210 N/mm$^2$, this differs from BS 5950: Part 1 which gives a value of 205 N/mm$^2$.

## 7.3  Section Classification

In C-EC3 the elements of cross-sections are classified as *Class 1* to *Class 4* as indicated in Chapter 2, Section 2.2 of this text. A more detailed, comprehensive treatment is given in C-EC3 than in BS 5950:Part 1. Three *Tables 5.6(a)*, *(b)* and *(c)* are given for classification of *Class 1*, *2* and *3* cross-sections respectively. The resistance of *Class 4* cross-sections is governed by local buckling. In BS 5950:Part 1 the strength of slender cross-sections is limited according to the requirements of *Clause 3.6* and the strength reduction factors in *Table 8*.

In C-EC 3 a more realistic approach is taken in which an effective width of section is used while still maintaining the full yield strength in resistance calculations. The SCI publication P115 gives the classification of standard sections under pure bending or pure compression in the Section Property Tables. The classification of flanges subject to combined bending and compression or tension is given in *Tables 5.6 (a)* to *(c)*. Webs subject to combined bending and compression or tension are dealt with in *Clause 5.3.2* and *Table 5.8* for *Class 1* and *Class 2* cross-sections and in *Table 5.9* for *Class 3* cross-sections.

## 7.4  Use of C-EC3

A selection of examples are used in this chapter to illustrate the use of C-EC3 for the design of structural elements. Where necessary, extracts from appropriate Tables in C-EC3 have been included, although access to a copy of C-EC3 is preferable. The clause reference numbers relate to those used in C-EC3 and **not** EC3 which differs slightly.

### 7.4.1  Example 7.1  Beam with full lateral restraint

A simply supported beam ABCD supports a uniformly distributed load and two point loads as shown in Figure 7.2. Using the design data given check the suitability of a 457 × 152 × 52 UB with respect to:

(i)     shear
(ii)    bending
(iii)   web crushing
(iv)    web crippling        (this is an additional check which is not required in BS 5950: Part 1),
(v)     web buckling        and
(vi)    deflection

Assume the compression flange to be fully restrained.

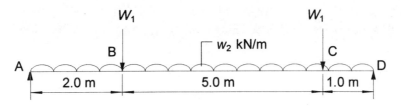

**Figure 7.2**

**Design Data:**
Point loads: $W_1$
Characteristic permanent load $\qquad\qquad\qquad\qquad G = 4.5$ kN
Characteristic variable load $\qquad\qquad\qquad\qquad Q = 22.5$ kN

Uniformly distributed load: $w_2$
Characteristic permanent UDL (includes self-weight) $\quad g = 1.4$ kN/m
Characteristic variable UDL $\qquad\qquad\qquad\qquad\quad q = 8.0$ kN/m

**Solution:**

Section properties: $457 \times 152 \times 52$ UB

| | | | |
|---|---|---|---|
| $h = 449{,}8$ mm | $b = 152{,}4$ mm | $t_w = 7{,}6$ mm | $t_f = 10{,}9$ mm |
| $c/t_f = 6{,}99$ | $d/t_w = 53{,}6$ | $d = 407{,}6$ mm | $i_{zz} = 3{,}11$ cm |
| $W_{el \cdot y} = 950$ cm$^3$ | $W_{pl \cdot y} = 1096$ cm$^3$ | $I_{yy} = 21370$ cm$^4$ | $A_v = 36{,}5$ cm$^2$ |

*Table 2*.1 Partial safety factors

| **Table 2.1** | Partial safety factors $\gamma_F$ - for nor⟨ |
|---|---|
| Dead load $\gamma_G$ | Imposed floor load $\gamma_Q$ |
| 1,35 | 1,5 |
| 1,35 | |
| 1,35 | |
| 1,35 | |

Extract from *Table 2.1 C-EC3 Concise Eurocode...*(The Steel Construction Institute)

Permanent load $\quad \gamma_G = 1{,}35$
Variable load $\qquad\; \gamma_Q = 1{,}5$
Factored point load $\;= (1{,}35 \times 4{,}5) + (1{,}5 \times 22{,}5) \quad = 39{,}8$ kN
Factored UDL $\qquad = (1{,}35 \times 1{,}4) + (1{,}5 \times 8{,}0) \quad\; = 13{,}9$ kN/m

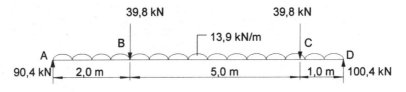

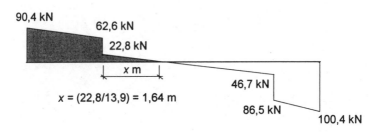

$$\text{Maximum bending moment } = \frac{(90,4+62,6)2}{2} + \frac{(1,64 \times 22,8)}{2} = 171,7 \text{ kNm}$$

| | | | |
|---|---|---|---|
| Design shear force | $V_{Sd}$ | = | 100,4 kN |
| Design bending moment | $M_{Sd}$ | = | 171,7 kNm |
| Coincident shear force | $V_{sd}$ | = | 62,6 kN |

Table 3.1   Material strength

| **Table 3.1** | Nominal yield strength $f_y$ and nominal ultimate tensile strength $f_u$ for steel | | |
|---|---|---|---|
| Nominal steel grade | | thickness | |
| BS EN 10025: 1990 | BS EN 10025: 1993 | $t \leq 40$ mm | |
| | | $f_y$(N/mm$^2$) | $f_y$(N/mm$^2$) |
| *Fe 430* | *S275* | 275 | 430 |

Extract from *Table 3.1 C-EC3 Concise Eurocode...*(The Steel Construction Institute)

*Grade S275* steel,        $t = 10,9 < 40$ mm

$$\therefore f_y = 275 \text{ N/mm}^2 \qquad f_u = 430 \text{ N/mm}^2$$

*Clause 5.3*        Section classification

*Table 5.6(a)*     Flange is subject to compression

| Table 5.6(a) | | Limiting width to thickness ratios for Class 1 elements | | | |
|---|---|---|---|---|---|
| Type of element | Loading | Stress distribution in element | Type of section | Width to thickness ratio | *Fe430 (S275)* |
| outstand element of compression flange | Flange subject to compression | | Welded | $c/t_f$ | $t \le 40$ mm 8,3 |
| | | | Rolled | $c/t_f$ | 9,2 |

Extract from *Table 5.6(a) C-EC3 Concise Eurocode...*(The Steel Construction Institute)

$$c/t_f = 6,99 \quad < \quad 9,2 \qquad \therefore \text{ Flange is Class 1}$$

Also from Table 5.6(a) Web is subject to bending

$$c/t_w = 53,6 \quad < \quad 66,6 \qquad \therefore \text{ Web is Class 1}$$

**Section is Class 1**

This classification is also given in the 'Properties of Structural Steel Sections', Part IV of SCI publication P115.

*Clause 5.5.1* Design plastic shear resistance

$$V_{pl \cdot Rd} = A_v \left( f_y / \sqrt{3} \right) / g_{M0}$$

The Shear Area is defined in *Table 5.16* and given in SCI publication P115 *Clause 5.1(1)*. For Class 1 section material partial safety factor $\gamma_{M0} = 1,05$

$$V_{pl \cdot Rd} = \frac{36,5 \times 10^2 \left( 275 / \sqrt{3} \right)}{1,05 \times 10^3} = 552 \text{ kN} \qquad V_{sd} < V_{pl \cdot Rd}$$

**Section is adequate in shear**

*Clause 5.5.2(1)* Design moment resistance

By inspection the coincident shear force $\ll 0,5 V_{pl \cdot Rd}$

*Clause 5.5.2.2(1)* For Class 1 sections

$$M_{c.Rd} = W_{pl} f_y / \gamma_{M0}$$

$$M_{c.Rd} = \frac{\left(1096 \times 10^3 \times 275\right)}{1,05 \times 10^6} = 287 \text{ kNm}$$

$$M_{sd} < M_{c.Rd}$$

*Clause 5.5.5(2)*      Since the compression flange is fully restrained there is no need to check for lateral torsional buckling.

**Section is adequate in bending**

*Clause 5.5.6.2*  Shear buckling

For *S275* grade steel the limiting value of    $d/t_w$   = 63,8
actual value of  $d/t_w$   = 53,6  < 63,8

**Shear buckling check is not required**

*Clause 5.7.1*   Resistance of web to transverse forces

In BS 5950:Part 1 web buckling (*Clause 4.5.2*) and web bearing (*Clause 4.5.3*) checks are carried out to determine if load bearing web stiffeners are required. In C-EC3 an additional check is also required, i.e. web crippling.

*Clause 5.7.3*  Web crushing     (i.e. bearing)

Design crushing resistance  $R_{y.Rd} = [(S_s + S_y)t_w f_{yw}]/\gamma_{M1}$

where:

$$S_y = t_f(b_f/t_w)^{0,5}[f_{yf}/f_{yw}]^{0,5}[1-(\gamma_{M0}\sigma_{f.Ed}/f_{yf})^2]^{0,5}$$

for an end bearing or point load at the end of a member      and

$$S_y = 2t_f(b_f/t_w)^{0,5}[f_{yf}/f_{yw}]^{0,5}[1-(\gamma_{M0}\sigma_{f.Ed}/f_{yf})^2]^{0,5}$$

for a bearing within the length of a member or a point load within the span.

$\sigma_{f.Ed}$   is the longitudinal stress in the flange
$b_f$    is the flange breadth b, but   $\leq 25t_f$
$S_s$    is the length of stiff bearing defined in *Figure 5.16* of C-EC3

Assume the value of $S_s$ to be 50 mm
Both the flanges and the web are *Class 1*        $\therefore f_{yf} = f_{yw} = 275 \text{ N/mm}^2$

*Clause 5.1(1)*       For Class 1 cross-sections   $\gamma_{M0}$   = 1,05
For buckling of sections    $\gamma_{M1}$   = 1,05

Check at support
At the end support the bending moment equals zero

$$\therefore \quad \sigma_{f.Ed} = \frac{M_{sd}}{\left(h-t_f\right)\left(h \times b\right)} = 0$$

check $b_f$ = 152,4 < (25 × 10,9)  o.k.

$$S_y = 10,9\left(\frac{152,4}{7,6}\right)^{0,5}\left(\frac{275}{275}\right)^{0,5}[1-0]^{0,5} = 48,81 \text{ mm}$$

$$R_{y.Rd} = [(48,81 + 50)7,6 \times 275]/(1,05 \times 10^3) = 148 \text{ kN}$$

Maximum reaction $F_{Sd}$ < $R_{y.Rd}$

**Web is adequate in bearing**

*Clause 5.7.4* Web crippling

Design crippling resistance $R_{a.Rd} = 0,5t_w^2(Ef_{yw})^{0,5}[(t_f/t_w)^{0,5} + 3(t_w/t_f)(S_s/d)]/\gamma_{M1}$

where:

$S_s$ is the stiff bearing length and $S_s/d \le 0,2$

In addition if a co-existent bending moment exists at the location where web crippling is being checked, the following conditions must be satisfied:

$$F_{Sd} \le R_{a.Rd}, \quad M_{Sd} \le M_{c.Rd}, \quad \text{and} \quad \frac{F_{Sd}}{R_{a.R}} + \frac{M_{Sd}}{M_{c.Rd}} \le 1,5$$

At support D

Assuming the stiff bearing length $\quad S_s = 50 \text{ mm}$

$$\frac{S_s}{d} = \frac{50}{407,6} = 0,12 < 0,2$$

*Clause 5.1(1)* $\qquad\qquad\qquad\qquad \gamma_{M1} = 1,05$

*Clause 3.1.4* Modulus of elasticity $\qquad E = 210 \times 10^3 \text{ N/mm}^2$

Design crippling resistance:

$$R_{a.Rd} = \frac{0,5 \times 7,6^2\left(210 \times 10^3 \times 275\right)^{0,5}}{1,05 \times 10^3}\left[\left(\frac{10,9}{7,6}\right)^{0,5} + 3\left(\frac{7,6}{10,9}\right)(0,12)\right] = 302 \text{ kN}$$

Maximum design shear force $\qquad V_{Sd} = 100,4 \text{ kN} < R_{a.Rd}$

**Web is adequate at supports**

At point B under point load

Under the point load the stiff bearing length will be at least equal to 50 mm and there is a coincident bending moment.

$$\therefore R_{a.Rd} \approx 302 \text{ kN}$$

*Clause    5.7.4(2)*                        $F_{sd}$   = 39,8 kN
                                            $R_{a.Rd}$ ≈ 302 kN
                                            $M_{sd}$   = 171,4 kNm
                                            $M_{c.Rd}$ = 287 kN

$$F_{Sd} \leq R_{a.Rd}, \quad M_{sd} \leq M_{c.Rd}, \quad \frac{F_{Sd}}{R_{a.Rd}} + \frac{M_{Sd}}{M_{c.Rd}} = \frac{39,8}{302} + \frac{171,4}{287} = 0,73 \leq 1,5$$

**Web is adequate under point load at point B**

By inspection web under the point load at C is adequate.

*Clause 5.7.5*          Web buckling
*Clause 5.4.3.2*        Design buckling resistance      $R_{b.Rd} = (\beta_A f_c A)/\gamma_{M1}$

where:
  $\beta_A = 1,0$
  $A = b_{eff} \times t_w$

$b_{eff}$  is the effective breadth of the web taken as $\left[ h^2 + S_s^2 \right]^{0,5}$. At the ends of a member

  $b_{eff}$ should not be taken as greater than the actual length available $\left( \dfrac{h}{2} + a \right)$,

  (where $a$ is the distance from the centre of support to the end of the beam), nor
  greater than '$h$'. This is indicated in *Clause 5.7.5(2)* and for various load cases in
  *Figure 5.17* of C-EC3.
$f_c$ =  is the compressive strength obtained from *Table 5.14 (a)* or *(b)* using buckling
  curve *(c)*

To obtain a value of $f_c$ from *Table 5.14* requires a value of slenderness ($\lambda$) which is
dependent on the conditions of lateral and rotational restraint of the flanges at the location
on the member where the force is applied.
Typical values of $\lambda$ for the more common situations are given in *Table 5.29*.

At support D
*Clause 5.7.5(4)* and *Table 5.29*  Assuming that the ends of the web are restrained against
both rotation and relative lateral movement.

$$\lambda = 2,5 d/t_w = 2,5 \times 53,6 = 134$$

*Table 5.13*      The web is considered to be a solid section        and

*Clause 5.7.5(3)*      ∴ Use *buckling curve c*

*Table 5.14(a)*          steel grade      *S275*        $\beta_A = 1,0$    $t \leq 40$ mm

| strength $f_c$ (N/mm$^2$) for steel grade *Fe 430* *(S275)* | | | | | | | | |
|---|---|---|---|---|---|---|---|---|
| | $t \le 40$ mm | | | | 40 mm $\le t \le 100$ mm | | | |
| $\lambda \sqrt{\beta_A}$ | Buckling curve | | | | Buckling curve | | | |
| | *a* | *b* | *c* | *d* | *a* | *b* | *c* | *d* |
| - | - | - | - | - | - | - | - | - |
| **130** | 103 | 94 | **87** | 76 | 102 | 93 | 85 | 75 |
| **135** | 96 | 89 | **82** | 72 | 95 | 88 | 80 | 71 |
| 140 | 90 | 84 | 77 | 68 | 90 | 83 | 76 | 67 |

Extract from *Table 5.14(a) C-EC3 Concise Eurocode...*(The Steel Construction Institute)

$$\lambda \sqrt{\beta_A} \quad = \quad 134 \qquad \therefore \quad f_c \ = \ 83 \ \text{N/mm}^2$$

*Clause 5.1(1)* $\qquad \gamma_{M1} \ = \ 1,05$

*Clause 5.7.5(1)* and *Table 5.17* $\quad$ assuming $\ a \ = \ 50$ mm

$$b_{eff} \quad = \quad \frac{449,2}{2} \ + 50 \quad = \ 274,6 \ \text{mm} \quad < \ h$$

*Clause 5.4.3.2* Design buckling resistance $\qquad R_{b.Rd} \ = \ (\beta_A f_c A)/\gamma_{M1}$

$$= \quad \frac{1,0 \times 83 \times (274,6 \times 7,6)}{1,05 \times 10^3}$$

$$= \quad 165 \ \text{kN}$$

Maximum design shear force $\qquad V_{Sd} \ = \ 100,4 \ \text{kN} \ < \ R_{b.Rd}$

**Web is adequate in buckling**

By inspection the web is adequate in buckling at points B and C.

*Clause 2.4*  Serviceability Limit State of Deflection

The check for the limit state of deflection differs from that used in BS 5950:Part 1. C-EC3 recommends limiting values of maximum vertical deflection due to variable loads and combined permanent and variable loads allowing for a pre-camber where it is used.

Second-order deflections induced by the rotational stiffness of semi-rigid joints and possible plastic deformations at the serviceability limit state should also be considered when appropriate.

The partial safety factors to be adopted when determining the serviceability design loads for deflection calculations are given in *Table 2.3*, they are:

(i)   permanent loads $\qquad \gamma_G \ = \ 1,0$
(ii)  variable floor loads $\quad \gamma_Q \ = \ 1,0$
(iii) variable roof loads $\quad \gamma_Q \ = \ 1,0$
(iv)  variable wind loads $\quad \gamma_Q \ = \ 0,9 \times 1,0 = 0,9$

For combinations of loading:

|  |  |  |  |  |  |  |  |  |  |
|---|---|---|---|---|---|---|---|---|---|
| (i) + (ii) | $\gamma_G$ | = | 1,0 | $\gamma_{Qfloor}$ | = | 1,0 | | | |
| (i) + (iii) | $\gamma_G$ | = | 1,0 | $\gamma_{Qroof}$ | = | 1,0 | | | |
| (i) + (iv) | $\gamma_G$ | = | 1,0 | $\gamma_{Qwind}$ | = | 0,9 | | | |
| (i) + (ii) + (iii) + (iv) | $\gamma_G$ | = | 1,0 | $\gamma_{Qfloor}$ | = | 1,0 | $\gamma_{Qroof}$ = 1,0 | $\gamma_{Qwind}$ = 0,81 | |

Design point loads:

| permanent load | $G_d$ = 1,0 × 1,0 | = 1,0 kN |
|---|---|---|
| variable | $Q_d$ = 1,0 × 5,0 | = 5,0 kN |

Design distributed loads:

| permanent load | $g_d$ = 1,0 × 1,4 | = 1,4 kN |
|---|---|---|
| variable load | $q_d$ = 1,0 × 8,0 | = 8,0 kN |

Since the beam has a non-standard load case, the deflection can be estimated using the 'equivalent distributed load' technique explained in section 2.6.1 of Chapter 2,   i.e.

$$\delta_{max} \approx \frac{0,104 \times \text{B.Max}L^2}{EI}$$

permanent loads:

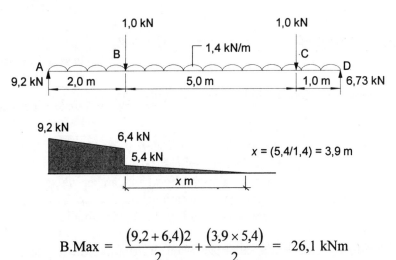

$$\text{B.Max} = \frac{(9,2+6,4)2}{2} + \frac{(3,9 \times 5,4)}{2} = 26,1 \text{ kNm}$$

variable loads:

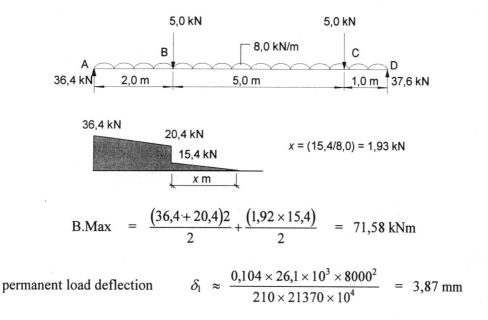

$$\text{B.Max} = \frac{(36,4 + 20,4)2}{2} + \frac{(1,92 \times 15,4)}{2} = 71,58 \text{ kNm}$$

permanent load deflection $\qquad \delta_1 \approx \dfrac{0,104 \times 26,1 \times 10^3 \times 8000^2}{210 \times 21370 \times 10^4} = 3,87 \text{ mm}$

variable load deflection $\qquad \delta_2 \approx 3,87 \times \dfrac{71,58}{26,1} = 10,6 \text{ mm}$

*Table 4.1*  Recommended limiting deflections:

due to variable loads $= \dfrac{L}{250} = \dfrac{8000}{250} = 32 \text{ mm}$

$$\delta_2 < 32 \text{ mm}$$

**Variable load deflection is acceptable**

*Figure 4.1* $\qquad\qquad \delta_{max} = \delta_1 + \delta_2 - \delta_0$

where $\delta_0 = $ pre-camber     In this the pre-camber equals zero

$$\delta_{max} = 3,87 + 10,6 = 14,47 \text{ mm}$$

*Table 4.1*  Recommended limiting deflections:

due to combined loads $\quad = \dfrac{L}{200} = \dfrac{8000}{200} = 40 \text{ mm}$

$$(\delta_1 + \delta_2) < 32 \text{ mm}$$

**Combined load deflection is acceptable**

**Lateral Torsional Buckling (LTB)   (*Clause 5.5.5*)**

*Clause 5.5.5(3)* indicates that lateral torsional buckling need not be checked if the spacing of restraints is such that:

$$\lambda_{LT}\sqrt{\beta_w} \quad \leq \quad 34,7 \qquad \text{for steel grade } S275$$
$$< \quad 30,6 \qquad \text{for steel grade } S355$$

where $\lambda_{LT}$ is the equivalent slenderness given in *Clause 5.5.5(a)* as:

$$\lambda_{LT} = \frac{\left[k / C_1\right]^{0,5} L / i_{LT}}{\left[1 + \dfrac{\left(L / a_{LT}\right)^2}{25,66}\right]^{0,25}}$$

where:

$k$     is the effective length factor from *Tables 5.21* and *5.24* for beams and cantilevers respectively.

$C_1$    is a factor to allow for the shape of the bending moment diagram and also depends upon the end restraint conditions. $C_1$ encompasses the equivalent uniform moment factor '$m$' and the slenderness correction factor '$n$' used in BS 5950:Part 1. The value of $C_1$ equals 1.0 for cantilevers, for other beams the values of the parameter $[k/c_1]^{0,5}$ are given in *Table 5.22* and *5.23* depending on the loading type.

$i_{LT}$ and $a_{LT}$ are as defined previously and given in Section Property Tables.

$\beta_w$   =   1,0 for *Class 1* or *Class 2* cross-sections

     =   $W_{el \cdot y}/W_{pl \cdot y}$ for *Class 3* cross-section

     =   $W_{eff \cdot y}/W_{pl \cdot y}$ for *Class 4* cross-sections as defined in *Clause 5.5.5(7)*

*Clause 5.5.5(4)*   The conditions of *Clause 5.5.5(3)* will be satisfied for *Class 1*, or *Class 2* rolled I, H or channel sections provided that:

$$\left[\frac{k}{C_1}\right]^{0,5} L \leq \quad 35,0 i_{LT} \quad \text{or } 37,5 i_z \text{ for steel grade } S275$$

$$\leq \quad 30,8 i_{LT} \quad \text{or } 33,0 i_z \text{ for steel grade } S355$$

*Clause 5.5.5(5)* and *Table 5.17*

The conditions of *Clause 5.5.5(3)* will be satisfied for rectangular hollow sections provided that the value of $\beta_w[k/C_1]L/i_z$ does not exceed the values given in *Table 5.17*. This is similar to *Clause B.2.6* and *Table 38* for boxed sections in *Appendix B* of BS 5950:Part 1.

     In *Clause 5.5.5(10)* an alternative equation for determining the equivalent slenderness $\lambda_{LT}$ is given:

$$\lambda_{LT} = \left[\frac{k}{C_1}\right]^{0,5} \lambda_{LTB}$$

where $\lambda_{LTB}$ is the basic equivalent slenderness obtained from *Table 5.19* using $L/i_{LT}$ and $a_{LT}/i_{LT}$.

In *Clause 5.5.5(11)* a further alternative is given to determine $f_b$ for a rolled section using *Table 5.20* and parameters $L / a_{LT}$ and $\beta_w^{0,5} \left[\dfrac{k}{C_1}\right]^{0,5} L/i_{LT}$

## Example 7.2 Beam with intermediate and end restraints

A single span beam ABCD of 8,0 m length supports point loads at B and C as shown in Figure 7.3. Using the design data given check the suitability of a $406 \times 140 \times 39$ UB with respect to lateral torsional buckling resistance.

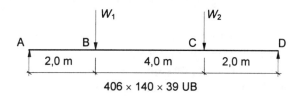

$$406 \times 140 \times 39 \text{ UB}$$

**Figure 7.3**

**Design Data:**
**Point load $W_1$**
| | |
|---|---|
| Characteristic permanent load | = 6,0 kN |
| Characteristic variable load | = 15,0 kN |

**Point load $W_2$**
| | |
|---|---|
| Characteristic permanent load | = 6,0 kN |
| Characteristic variable load | = 24,0 kN |

**Solution:**
Section properties: $457 \times 152 \times 52$ UB

| | | | |
|---|---|---|---|
| $h$ = 397,3 mm | $b$ = 141,8 mm | $t_w$ = 6,3 mm | $t_f$ = 8,6 mm |
| $c/t_f$ = 8,24 | $d/t_w$ = 57,1 | $d$ = 359,7 mm | $i_{zz}$ = 2,89 mm |
| $W_{el\cdot y}$ = 625 cm³ | $W_{pl\cdot y}$ = 718 cm³ | $I_{yy}$ = 12410 cm⁴ | $A_v$ = 27,1 cm² |
| $a_{LT}/i_{LT}$ = 36,3 | $i_{LT}$ = 3,33 cm | $a_{LT}$ = 12,1 cm | |

EC3 Section classification = 1.0

*Table 2.1* Partial safety factors
| | | |
|---|---|---|
| | Permanent loads | $\gamma_G$ = 1,35 |
| | Variable loads | $\gamma_Q$ = 1,5 |

Factored load    $W_1 = [(1,35 \times 6) + (1,5 \times 15)] = 30,6 \text{ kN}$
Factored load    $W_2 = [(1,35 \times 6) + (1,5 \times 24)] = 44,1 \text{ kN}$

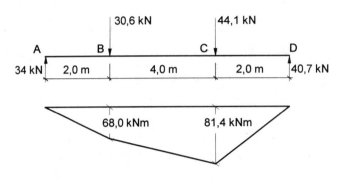

*Table 3.1*    Material strength    Grade *S275* steel    $t = 6,3 \text{ mm} < 40 \text{ mm}$
$f_y = 275 \text{ N/mm}^2$    $f_u = 430 \text{ N/mm}^2$

---

*Clause 5.3*    Section classification: This is given in the section property tables and is shown here for illustration purposes.

*Table 5.6*    Flange subject to compression
$c/t_f = 8,24 < 9,2$    Flange is Class 1
Web is subject to bending
$d/t_w = 57,1 < 66,6$    Web is Class 1

**Section is Class 1**

---

**Check the critical length between B and C**

Assumptions    compressive flange is laterally restrained at A, B, C and D
beam is fully restrained against torsion
both flanges are free to rotate on plan

*Table 5.21*    $k = 1,0$

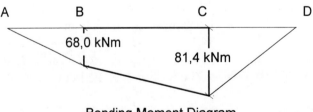

Bending Moment Diagram

$$\psi = \frac{\text{smaller end moment}}{\text{larger end moment}} = \frac{68,0}{81,4} = 0,835$$

*Table 5.22* Factors $[k/C_1]^{0,5}$ for end moment loading with $k = 1,0$

| Table 5.22 | Factors $[k/C_1]^{0,5}$ and $[k/C_3]$ | | : End moment loading | | | |
|---|---|---|---|---|---|---|
| $M$ $\qquad \psi M$ | $k = 1,0$ | | $k = 0,85$ | | $k = 0,7$ |
| $\psi$ | $[k/C_1]^{0,5}$ | $[k/C_3]$ | $[k/C_1]^{0,5}$ | $[k/C_3]$ | $[k/C_1]^{0,5}$ |
| - | - | - | - | - | - |
| +0,9 | 0,97 | 1,00 | 0,89 | 0,70 | 0,79 |
| +0,8 | 0,95 | 1,00 | 0,85 | 0,67 | 0,76 |
| +0,7 | 0,92 | 1,00 | 0,83 | 0,66 | 0,73 |

Extract from Table 5.22 *C-EC3 Concise Eurocode...*(The Steel Construction Institute)

By interpolation $\quad [k/C_1]^{0,5} = 0,957$

*Clause 5.5.5(4)*:

$$\left[\frac{k}{C_1}\right]^{0,5} L = 0,957 \times 4000 = 3828 \text{ mm}, \quad (35,0 i_{LT}) = 35,0 \times 33,3 = 1166 \text{ mm}$$

$$\left[\frac{k}{C_1}\right]^{0,5} L > 35,0 i_{LT} \qquad \therefore \textbf{ Length BC of the beam must be checked for LTB}$$

*Clause 5.5.5(7)* $\qquad M_{b.Rd} = \dfrac{\beta_w f_b W_{pl.y}}{\gamma_{M1}}$

where $f_b$ is the bending strength and must be determined from *Table 5.18(a)* for rolled sections using $f_y$ and the modified slenderness $\quad \lambda\sqrt{\beta_w}$

$$\lambda_{LT} = \frac{[k/C_1]^{0,5} L/i_{LT}}{\left[1+\dfrac{(L/a_{LT})^2}{25,66}\right]^{0,25}} = \frac{0,957 \times (4000/33,3)}{\left[1+\dfrac{(4000/121)^2}{25,66}\right]^{0,25}} = \frac{114,95}{2,57} = 44,73$$

$\beta_w = 1,0$ for Class 1 cross-sections

*Table 5.18(a)* $\qquad \lambda_{LT}\sqrt{\beta_w} = 44,73 \qquad$ and $\qquad f_y = 275 \text{ N/mm}^2$

| Table 5.18(a) | Bending strength $f_b$ (N/mm$^2$) = Rolled | | | |
|---|---|---|---|---|
| $\lambda_{LT}\sqrt{\beta_W}$ | Yield strength $f_y$ (N/mm$^2$) | | | |
| | 255 | **275** | 335 | 355 |
| - | - | - | - | - |
| **44** | 237 | **254** | 303 | 319 |
| **46** | 235 | **252** | 300 | 316 |

Extract from *Table 5.18(a) C-EC3 Concise Eurocode*...(The Steel Construction Institute)

$$f_b = \textbf{253 N/mm}^2$$

*Clause 5.1(1)* $\qquad \gamma_{M1} = 1,05$

$$M_{b.Rd} = \frac{\beta_w f_b W_{pl.y}}{\gamma_{M1}} = \frac{1,0 \times 253 \times 718 \times 10^3}{1,05 \times 10^6} = 173 \text{ kNm}$$

Maximum design moment between B and C $\qquad M_{Sd} = 81,4 \text{ kNm}$

$$M_{Sd} < M_{b.Rd}$$

By inspection the lengths of beam between AB and BC are adequate.

**Section is adequate w.r.t LTB**

**Alternative method of evaluating equivalent slenderness $\lambda_{LT}$**

*Clause 5.5.5(10)*

*Table 5.19* $\qquad L/i_{LT} = \dfrac{4000}{33,3} = 120$

$$a_{LT}/i_{LT} = \frac{121}{33,3} = 36,3 \quad \text{(or from section property tables)}$$

| Table 5.19 | Basic equivalent slenderness $\lambda_{LTB}$ | | | | | | | | | |
|---|---|---|---|---|---|---|---|---|---|---|
| $L/i_{LT}$ | $a_{LT}/i_{LT}$ | | | | | | | | | |
| | 4 | 6 | 8 | 10 | 15 | 20 | 25 | 30 | **35** | **40** |
| - | - | - | - | - | - | - | - | - | - | - |
| 115 | 47,9 | 58,1 | 66,3 | 73,0 | 85,4 | 93,5 | 98,9 | 102,7 | 105,3 | 107,2 |
| **120** | 49,0 | 59,5 | 67,9 | 74,8 | 87,8 | 96,4 | 102,2 | 106,3 | **109,2** | **111,3** |
| 125 | 50,0 | 60,8 | 69,4 | 76,6 | 90,1 | 99,2 | 105,5 | 109,9 | 113,0 | 115,3 |

Extract from *Table 5.19 C-EC3 Concise Eurocode*...(The Steel Construction Institute)

$$\lambda_{LTB} = \textbf{109,75}$$

$$\lambda_{LT} = \left[\frac{k}{C_1}\right]^{0,5} \lambda_{LTB} = 0,975 \times 109,75 = 104$$

*Table 5.18(a)* $\quad \lambda_{LT}\sqrt{\beta_w} = 109,75 \times 1,0 = 109,75, \quad f_y = 275 \text{ N/mm}^2$

$$f_b = \mathbf{146 \text{ N/mm}^2}$$

the value of $M_{b.Rd}$ can be determined as before.

## Alternative method of evaluating bending strength $f_b$

*Clause 5.5.5(11)*

$$L/a_{LT} = \frac{4000}{121} = 33,1 \qquad \beta_w = 1,0$$

$$\beta_w^{0,5}\left[\frac{k}{C_1}\right]^{0,5} L/i_{LT} = \frac{1,0 \times 0,957 \times 4000}{33,3} = 115$$

| **Table 5.20(a)** Bending strength $f_b$ for ROLLED SECTIONS Grade *Fe 430* (*S275*) steel, $t \le 40$ mm ($f_y = 275$ N/mm²) | | | | | | | | | |
|---|---|---|---|---|---|---|---|---|---|
| $\beta W^{0,5}[k/C_1]^{0,5}L/i_{LT}$ | $L/a_{LT}$ | | | | | | | | |
| | 0 | 2 | 4 | 6 | - | - | - | 30 | 35 |
| - | - | - | - | - | - | - | - | - | - |
| 110 | 135 | 142 | 160 | 180 | - | - | - | 253 | 256 |
| **115** | 126 | 133 | 150 | 171 | - | - | - | **251** | 254 |
| 120 | 117 | 124 | 142 | 162 | - | - | - | 248 | 252 |

Extract from *Table 5.20(a) C-EC3 Concise Eurocode...*(The Steel Construction Institute)

$$f_b = \mathbf{253 \text{ N/mm}^2}$$

The value of $M_{b.Rd}$ can be determined as before.

## 7.7 Example 7.3 Truss members with axial tension/compression

An N-girder supports characteristic permanent and variable loads as shown in     Figure 7.4. Using the design data provided check the suitability of the sections indicated for members AB and CD.

**Design Data:**
**Point load $W_1$**
Characteristic permanent load        15 kN
Characteristic variable load          30 kN

**Point load   $W_2$**

| | |
|---|---:|
| Characteristic permanent load | 7,5 kN |
| Characteristic variable load | 15 kN |

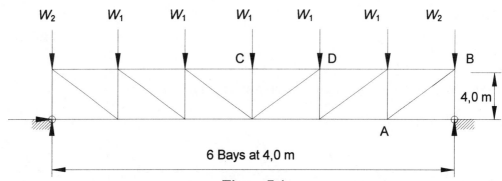

**Figure 7.4**

Member     AB:   $100 \times 100 \times 10$ single angle, double bolted to a gusset plate

Member     CD:   compound strut comprising $2/150 \times 90 \times 12$ angles double bolted with long legs connected to both sides of a gusset plate.

**Note:**   Assume 20 mm diameter bolts in 2 mm clearance holes at a pitch of $2,5 \times$ bolt diameter for all connections

**Solution:**

*Table 2.1*   Partial safety factors        $\gamma_G = 1,35$        $\gamma_Q = 1,5$

Factored load        $W_1 = (1,35 \times 15) + (1,5 \times 30) = 65,25$ kN
Factored load        $W_2 = (1,35 \times 7,5) + (1,5 \times 15) = 32,63$ kN

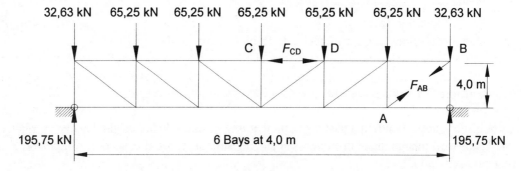

**Figure 7.5**

Member force AB $(N_{Sd}) = +230,7$ kN        tension
Member force CD $(N_{Sd}) = -293,6$ kN        compression

*Table 3.1* Material strength Grade *S275* steel, $t \leq 40$ mm
$f_y = 275$ N/mm² $f_u = 430$ N/mm²

**Member AB:** $100 \times 100 \times 12$ single angle, double bolted to gusset plate
Section properties: $A = 22.8$ cm²

*Clause 5.4.2(1)* Tension Members
Design tension resistance $= N_{Sd} \leq N_{t.Rd}$

*Clause 5.4.2(2)*
where:
$$N_{t.Rd} \leq N_{pl.Rd} = Af_y/\gamma_{M0}$$
$$N_{t.Rd} \leq N_{u.Rd} = 0.9A_{net}f_u/\gamma_{M2}$$

*Clause 5.4.2(6)* For angles connected through one leg reference is made to
*Clauses 5.8.3.2* for bolted connections and *Clause 5.8.3.3* for welded connections.

*Clause 5.8.3.2* As in BS 5950:Part 1 allowance is made for secondary effects due to eccentric connections in angles. In C-EC3, for angles with a single row of bolts in one leg the design ultimate resistance of the net section is determined from:

single-bolted $\quad N_{u.Rd} = \dfrac{2.0(e_2 - 0.5d_o)tf_u}{\gamma_{M2}}$

double-bolted $\quad N_{u.Rd} = \dfrac{\beta_2 A_{net} f_u}{\gamma_{M2}}$

more than two bolts $\quad N_{u.Rd} = \dfrac{\beta_3 A_{net} f_u}{\gamma_{M2}}$

where:
$d_o$ is the diameter of the bolt hole
$e_2$ is the edge distance of the bolt hole
$\beta_2$ is the reduction factor for 2 bolts as given in Table 5.30
$\beta_3$ is the reduction factor for 3 bolts as given in Table 5.30
$A_{net}$ is the net cross-sectional area of the angle. For an unequal angle connected by it's smaller leg, $A_{net}$ must be taken as the net cross-sectional area of an equal of leg size equal to the smaller leg.

**Note:** This differs from BS 5950:Part 1 in that it applies to **all** bolted angles connected by one leg i.e. both tension **and** compression members.

*Clause 5.1(1)* Material partial safety factors $\quad \gamma_{M0} = 1.05 \quad \gamma_{M2} = 1.2$

*Clause 5.4.2(2)* $\quad N_{pl.Rd} = Af_y/\gamma_{M0} \quad = \dfrac{(22.8 \times 10^2 \times 275)}{1.05 \times 10^3} = 597$ kN

*Clause 5.8.3.2*
$$N_{u.Rd} = \frac{\beta_2 A_{net} f_u}{\gamma_{M2}}$$

*Table 5.30*

| Table 5.30 | Reduction factors $\beta_2$ and | |
|---|---|---|
| Pitch $p_1$ | $p_1 \leq 2,5\, d_o$ | $2,5 < p_1 < 5,0\, d_o$ |
| 2 bolts | $\beta_2 = 0,4$ | $\beta_2 = 0,1 + 1,2 p_1$ |
| $\geq 3$ bolts | $\beta_3 = 0,5$ | $\beta_3 = 0,3 + 0,8 p_1$ |

Extract from *Table 5.30 C-EC3 Concise Eurocode...*(The Steel Construction Institute)

pitch of bolts $= 2,5 \times 20 = 50$ mm $< 2,5 d_o$
For 2 bolts $\beta_2 = 0,4$

*Clause 5.4.1.2(1)*
$$A_{net} = [A_{gross} - (d_o \times t)] = [2280 - (22 \times 12)] = 2016 \text{ mm}^2$$

*Clause 5.8.3.2* $\quad N_{u.Rd} = \dfrac{\beta_2 A_{net} f_u}{\gamma_{M2}} = \dfrac{0,4 \times 2016 \times 430}{1,2 \times 10^3} = 289$ kN

$\therefore N_{t.Rd} = 289$ kN
$N_{Sd} < N_{t.Rd}$

**Section is adequate**

**Member CD:** 2/150 × 90 × 12 double angles, double bolted to both side of a 12 mm gusset plate

Section properties:
2/150 × 90 × 12 double angles
$\quad\quad A = 55,2 \text{ cm}^2 \quad i_{zz} = 3,7 \text{ cm} \quad i_{yy} = 4,78 \text{ cm}$
1/150 × 90 × 12 single angle
$\quad\quad A = 27,6 \text{ cm}^2 \quad i_{yy} = 4,78 \text{ cm} \quad i_{zz} = 2,5 \text{ cm} \quad i_{vv} = 1,95 \text{ cm}$

In C-EC3 the length of the longer leg is denoted by '$h$',and the shorter leg by '$b$'.
As in BS 5950:Part 1 the thickness is denoted by '$t$'.

Section Classification: *Table 5.6(c)* $\quad \dfrac{h}{t} = \dfrac{150}{12} = 12,5 < 13,9$

$$\frac{b+h}{t} = \frac{150+90}{12} = 20,0 < 21,3$$

**Section is Class 3**

*Clause 5.4.3(1)*     Compression Members

$$N_{Sd} \quad \le \quad N_{c.Rd} \qquad \text{at every cross-section}$$
$$\le \quad N_{b.Rd} \qquad \text{for the member as a whole}$$

where:

$N_{c.Rd}$  is the design compression resistance of the cross-section

$N_{b.Rd}$  is the design buckling resistance of the member given by *Clause 5.4.3.2(1)* as:

$$N_{b.Rd} \quad = \quad \beta_A f_c A / \gamma_{M1}$$

$\beta_A =$  1,0 for Class 1,2 or 3 cross-sections

$\quad =$  $A_{eff}/A$ for Class 4 cross-sections where $A_{eff}$ is defined in *Clause 5.3.4*

$\quad f_c$  is the compressive strength obtained from the appropriate buckling curve in *Table 5.14* for the relevant value of the modified slenderness ration $\lambda\sqrt{\beta_A}$

For angles connected through one leg the in-plane slenderness ratios are given in *Clause 5.8.3.1*. These values also apply to back-to-back double angles with the exception of the check for the v-v axis which is unnecessary.

*Clause 5.8.3.1*  For *S275* grade steel;

$$\lambda_{eff.v} \quad = \quad 0,7\lambda_v + 30,4 \qquad \text{(not required for double angles)}$$
$$\lambda_{eff.y} \quad = \quad 0,7\lambda_y + 43,4$$
$$\lambda_{eff.z} \quad = \quad 0,7\lambda_z + 43,4$$

The effective slenderness is used with the buckling curve in *Table 5.4*

Assuming the double angles are held apart by packing pieces 450 mm from each end and at 475 mm centres along the length.
Consider the combined double angle buckling over the full length.

For the double angle:  $L_e$ = 4000 mm   $\lambda_{yy}$  $= \dfrac{4000}{47,8} = 83,7$

$$\lambda_{zz} \quad = \quad \dfrac{4000}{37,0} = 108,1$$

Consider each component buckling between the packing pieces.

For the single angle:   $L_e$ = 475 mm     $\lambda_{vv} = \dfrac{475}{19,5} = 24$

$$\lambda_{yy} \quad = \quad \dfrac{475}{47,8} = 9.9$$

$$\lambda_{zz} \; = \; \frac{475}{25} \; = \; 19$$

*Clause 5.8.3.1*    By inspection $\lambda_{\text{eff.}z}$ is the critical value

*Clause 5.4.3.2(5)*    $\lambda_{\text{eff.}z} \; = \; (0,7 \times 108,1) + 43,4 \; = \; 119,1 \quad < \; 180$

*Table 5.14(a)*    Buckling curve c    $t \leq 40$ mm    $f_c = \; \mathbf{99,0 \; N/mm^2}$

*Clause 5.1(1)*    $\gamma_{M1} \; = \; 1,05 \quad \gamma_{M2} = \; 1,2 \quad \beta_A \; = \; 1,0$

*Clause 5.4.3.2(1)*    $N_{b.Rd} \; = \; \dfrac{\beta_A f_c A}{\gamma_{M1}} \quad = \; \dfrac{1,0 \times 99 \times 55,2 \times 10^2}{1,05 \times 10^3} \; = \; 520 \; kN$

*Clause 5.4.3.1(2)*    $N_{c.Rd} \; = \; \dfrac{A f_y}{\gamma_{M0}} \quad = \; \dfrac{55,2 \times 10^2 \times 275}{1,05 \times 10^3} \; = \; 1446 \; kN$

*Clause 5.8.3.2(2)*    $N_{u.Rd} \; = \; \dfrac{\beta_2 A_{net} f_u}{\gamma_{M2}}$

$A_{net} \; = \; [5520 - (22 \times 12)] = \; 5256 \; mm^2$

*Table 5.30*    $\beta_2 \; = \; 0,4$

$N_{u.Rd} \; = \; \dfrac{0,4 \times 5256 \times 430}{1,2 \times 10^3} \; = \; 753 \; kN$

$\therefore \; N_{c.Rd} \; = \; 520 \; kN$

$N_{Sd} \; < \; N_{c.Rd}$                          **Section is adequate**

## 7.8  Example 7.4  Concentrically loaded single storey column

A single storey column supports a symmetrical arrangement of beams as shown in Figure 7.6. Using the data provided check the suitability of a 203 × 203 × 60 UC section and design a suitable slab baseplate.

**Design Loads:**
$F_1$ = characteristic dead load        $G_k$ = 75 kN
$F_1$ = characteristic imposed load      $Q_k$ = 175 kN
$F_2$ = characteristic dead load        $G_k$ = 20 kN
$F_2$ = characteristic imposed load      $Q_k$ = 75 kN

Assume a characteristic concrete cube strength $f_{ck} = 30 \text{ N/mm}^2$

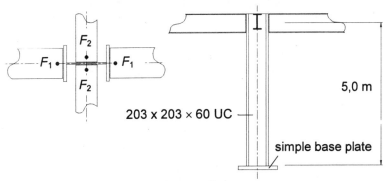

$203 \times 203 \times 60 \text{ UC}$

5,0 m

simple base plate

**Figure 7.6**

**Solution:**

Section properties: $203 \times 203 \times 60$ UC

| | | | |
|---|---|---|---|
| $h = 209,6$ mm | $b = 205,2$ mm | $t_w = 9,3$ mm | $t_f = 14,2$ mm |
| $c/t_f = 7,23$ | $d/t_w = 17,3$ | $d = 160,8$ mm | $i_{zz} = 5,19$ mm |
| $i_{yy} = 9,0$ cm | $i_{zz} = 5,19$ cm | $A = 76,0 \text{ cm}^2$ | |

EC3 Section classification $= 1.0$

*Table 2.1*    Partial Safety Factors

| | | | |
|---|---|---|---|
| Permanent loads | $\gamma_G = 1,35$ | | |
| Variable loads | $\gamma_Q = 1,5$ | | |
| Factored load $W_1$ | $= [(1,35 \times 75) + (1,5 \times 150)]$ | $= 326,3$ kN |
| Factored load $W_2$ | $= [(1,35 \times 20) + (1,5 \times 65)]$ | $= 124,5$ kN |
| Applied axial load | $N_{sd} = 2(W_1 + W_2) = 2(326,3 + 124,5)$ | $\approx 902$ kN |

*Table 3.1*    Material strength        Grade *S275* steel      $t < 40$ mm

$\qquad\qquad\qquad\qquad\qquad\qquad f_y = 275 \text{ N/mm}^2 \qquad f_u = 430 \text{ N/mm}^2$

*Clause 5.1(1)* $\qquad\qquad \gamma_{M0} = 1,05 \qquad \gamma_{M1} = 1,05 \qquad \gamma_{M2} = 1,2$

*Clause 5.4.3.1(1)* $\qquad\qquad N_{Sd} \leq N_{c.Rd}$

*Clause 5.4.3.1(2)* $\qquad\qquad$ For a Class 1 section $\qquad\qquad N_{c.Rd} = \dfrac{A f_y}{\gamma_{M0}}$

$$N_{c.Rd} = \frac{7600 \times 275}{1,05 \times 10^3}$$

$$N_{Sd} < N_{c.Rd}$$

**Section is adequate in compression**

*Clause 5.4.3.2.(1)* $\qquad\qquad N_{b.Rd} = \beta_A f_c A / \gamma_{M1} \qquad\qquad$ for Class 1 sections $\quad \beta_A = 1,0$

Assume the column is pinned at the top and the bottom

*Table 5.12*        Effective buckling lengths for various end restraint conditions

Effective buckling length about yy axis    $= l_{yy} = 1,0 \times 5000 = 5000$ mm
Effective buckling length about zz axis    $= l_{zz} = 1,0 \times 5000 = 5000$ mm

*Clause 5.4.3.2(2)*     $\lambda_{yy} = \dfrac{l_{yy}}{i_{yy}} = \dfrac{5000}{9,0} = 55,6$

$\lambda_{zz} = \dfrac{l_{zz}}{i_{zz}} = \dfrac{5000}{51,9} = 96,3$

*Clause 5.4.3.2(5)*     $\lambda_{yy} < 180$        $\lambda_{zz} < 180$

*Table 5.13*        Selection of appropriate buckling curve depending on type of cross-section and *h/b* ratio
For y-y axis  use *curve b*,    for z-z axis use *curve c*

*Table 5.14(a)*    Grade *S275* steel
For y-y axis            $\lambda\sqrt{\beta_A} = 55,6$        $f_c = 210$ N/mm$^2$
For z-z axis            $\lambda\sqrt{\beta_A} = 96,3$        $f_c = 131$ N/mm$^2$
**Critical value of $f_c$= 131 N/mm$^2$**

$$N_{b.Rd} = \frac{\beta_A f_c A}{\gamma_{M1}} = \frac{1,0 \times 131 \times 7600}{1,05 \times 10^3} = 948 \text{ kN}$$

$$N_{sd} < N_{b.Rd}$$

**Section is adequate**

*Clause 6.10*    Design of slab baseplate
    The design of column slab baseplates is more comprehensive and detailed than the use of the empirical equation given in *Clause 4.13.2* of BS 5950:Part 1. The minimum thickness, which should not be less than the flange thickness of the column being supported, is related to an effective projection 'c' of the baseplate from the face of the column or any stiffeners as shown in *Figures 6.18* and *6.19* of C-EC3. This is illustrated in this example.

*Clause 6.10.1(3)*    The effective projection should be determined from:

$$c = t \left[ \frac{f_y}{3f_j \gamma_{M0}} \right]^{0,5}$$

where:

$t$      is the thickness of the baseplate

$f_y$      is the yield strength of the baseplate material

$f_j$      is the bearing strength of the joint given by   $f_j = \beta_j k_j f_{cd}$

     where:

     $f_{cd}$ is the design value of the concrete compressive strength from *Table 6.12*

     $\beta_j$ is the joint coefficient. This can be taken as 2/3 provided that:

         (i) the characteristic strength of the grout is not less than $0{,}2 \times$ characteristic strength of the concrete ($f_{ck.cube}$)

         (ii) the thickness of the grout does not exceed $0{,}2 \times$ the smallest plan dimension of the baseplate

     $k_j$ is the concentration factor which is given by    $k_j = \left[ \dfrac{a_1 b_1}{ab} \right]^{0,5}$

where '$a$' and '$b$' are the length and width of the baseplate respectively and '$a_1$' and '$b_1$' are the length and width of the effective area of concrete foundation as given in *Figure 6.20* of C-EC3.

     Alternatively a conservative value of $k_j$ can be taken as equal to 1,0.

Assume the characteristic strength of the grout is the same as the characteristic strength of the concrete

|  |  |
|---|---|
|  | $f_{ck} = $ C30   and   $f_{gk} = $ 30 N/mm$^2$ |
| *Table 6.12* | $f_{cd} = f_{ck}/\gamma_c$         In Eurocode 2 [EC2] for concrete design the value of   $\gamma_c = 1{,}5$ |
| *Clause 6.10.1(5)* | $f_{cd} = \dfrac{30}{1{,}5} = 20$ N/mm$^2$ |
| *Clause 6.10.1(6)* | $f_{gk} \geq 0{,}2 \times f_{ck}$         $\therefore \beta_j = 2/3$ |
| *Clause 6.10.1(7)* | Assume $k_j = 1{,}0$ |
| *Clause 6.10.1(5)* | $f_j = \beta_j k_j f_{cd} = (0{,}67 \times 1{,}0 \times 20) = 13{,}4$ N/mm$^2$ |
| *Table 3.1* | Grade *S275* steel         $f_y = 275$ N/mm$^2$ |
| *Clause 5.1(1)* | $\gamma_{M0} = 1{,}05$ |

Assume   $t \approx$ the column flange thickness      $\therefore t = 15$ mm

*Clause 6.10.1(3)*      Effective projection   '$c$'

$$c = t \left[ \frac{f_y}{3 f_j \gamma_{M0}} \right]^{0,5} = 15 \left[ \frac{275}{3 \times 13{,}4 \times 1{,}05} \right]^{0,5} = 38.3 \text{ mm}$$

This effective projection should be considered all around each element of the column including any baseplate stiffeners.

Try baseplate   $\approx$   $(h + 2c) \times (h + 2c) \times 15$ mm thick.

$= [209{,}6 + (2 \times 38{,}3)]$   $= 286{,}2$ mm          say 290 mm square

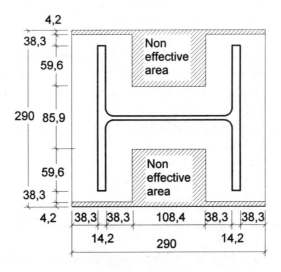

**Figure 7.7**

Effective bearing area $= (290 \times 290) - 2[(290 \times 4{,}1) + (108{,}4 \times 59{,}65)]$

$= 68{,}79 \times 10^3$ mm$^2$

Effective bearing pressure $= \dfrac{902 \times 10^3}{68{,}79 \times 10^3} = 13.23$ N/mm$^2$   $< f_j$

**Adopt 290 × 290 × 15 mm thick baseplate**

## 7.9  Axially Loaded Members with Moments

The interaction between bending and axial loading is discussed in Chapter 4 and a number of interaction equations given which are adopted by BS 5950:Part 1. These equations allow for the   secondary effects caused by one type of loading on  member behaviour which is subject to another type of loading.

C-EC3, *Clause 5.6* also provides interaction equations to take account of secondary effects. Three cases are considered.

(1)    Resistance of cross-section - (re local capacity check in BS 5950:Part 1)

*Clause 5.6.1.2*
For Class 1 or 2 cross-sections with low shear

$$M_{Sd} \;\leq\; M_{N.Rd}$$

where:    $M_{N.Rd}$ is the reduced plastic resistance moment allowing for an axial force $N_{sd}$. The value of $M_{N.Rd}$ can be determined using the approximate expression from *Table 5.27*. The U.K. NAD also permits the use of the relevant expressions given in U.K. section property tables as an alternative to *Table 5.27*.

As with BS 5950:Part 1 both a conservative and a more exact expression is given e.g. Assuming 'low shear' for bi-axial bending the Conservative Approach in *Clause 5.6.1.2(3)* uses the expression:

$$\frac{N_{Sd}}{Af_y / \gamma_{M0}} + \frac{M_{y.Sd}}{M_{pl.y.Rd}} + \frac{M_{z.Sd}}{M_{pl.z.Rd}} \leq 1.0$$

The More Exact Approach in *Clause 5.6.1.2(4)* uses the expression:

$$\left[\frac{M_{y.Sd}}{M_{Ny.Rd}}\right]^\alpha + \left[\frac{M_{z.Sd}}{M_{Nz.Rd}}\right]^\beta \leq 1.0$$

in which $\alpha$ and $\beta$ are given in *Table 5.28*.
Similar expressions to the Conservative Approach are given for Class 3 and Class 4 cross-sections.

C-EC3, *Clause 5.6.1.5* gives criteria when 'high shear' is present

(2)    Buckling Resistance of members with combined bending and axial tension.
*Clause 5.6.2(1)*
When members are susceptible to lateral torsional buckling, i.e.( $\lambda_{LT}\sqrt{\beta_W}$ > 34,7 for grade *S275* steel and 30,6 for grade *S355* steel), then

$$M_{eff.Sd} \leq M_{b.Rd}$$

where:

$M_{N.Rd}$    is the design buckling resistance moment = $\dfrac{\beta_W f_b W_{pl.y}}{\gamma_{M1}}$

$M_{eff.Sd}$    is the effective internal moment = $W_{com}\sigma_{com.Ed}$
where:
$W_{com.Ed}$    is the elastic section modulus for the extreme compression fibre

$S_{com.Ed}$    $= \left[\dfrac{M_{Sd}}{W_{com}} - \dfrac{\psi_{vec}N_{t.Sd}}{A}\right]$

$\psi_{vec}$    = 0,7    in the U.K. NAD
$N_{t.Sd}$    = design value of axial tension.

(3)    Buckling Resistance of members with combined bending and axial compression
*Clause 5.6.3(1)*
Three conditions are considered in *Clause 5.6.3(1)*

(a)    Axial compression and major axis bending

$$\frac{N_{Sd}}{N_{b.min.Rd}} + \frac{k_y M_{y.Sd}}{\eta M_{c.y.Rd}} \leq 1,0$$

where:

$N_{b.min.Rd}$   is the minimum design buckling resistance moment $= \dfrac{\beta_W f_b A}{\gamma_{M1}}$

The use of $N_{b.min.Rd}$ instead of $N_{b.y.Rd}$ as given in C-EC3 is conservative

$M_{c.y.Rd}$   is the design moment resistance for major axis bending.

$\qquad = \dfrac{W_{pl} f_y}{\gamma_{M0}}$   for Class 1 or Class 2 cross-sections

$\qquad = \dfrac{W_{el} f_y}{\gamma_{M0}}$   for Class 3 cross-sections

$\qquad = \dfrac{W_{eff} f_y}{\gamma_{M1}}$   for Class 1 0r Class 2 cross-sections

$k_y$   $= 1,5$   this is a conservative value, a procedure is given in C-EC3
$\qquad\qquad\qquad$ for determining a more precise value,

$\eta$   $= \dfrac{\gamma_{M0}}{\gamma_{M1}}$   for Class 1,2 and 3 cross-sections

$\qquad = 1,0$   for Class 4 cross-sections

For members which are susceptible to LTB *

$$\left[ \frac{N_{Sd}}{N_{b.z.Rd}} + \frac{k_{LT} M_{y.Sd}}{M_{b.Rd}} \right] \leq 1,0$$

$N_{b.z.Rd}$ $\qquad$ is the design buckling resistance for the minor axis.

$M_{b.Rd}$ $\qquad$ is the design buckling resistance moment $= \dfrac{\beta_W f_b W_{pl.y}}{\gamma_{M1}}$

$k_{LT}$ $\qquad = 1,0$   this is a conservative value, a procedure is given in C-EC3
$\qquad\qquad\qquad$ for determining a more precise value,

Additional moments must be added for Class 4 cross-sections.

(b)     Axial compression and minor axis bending

$$\frac{N_{Sd}}{N_{b.min.Rd}} + \frac{k_y M_{z.Sd}}{\eta M_{c.z.Rd}} \quad \leq \quad 1,0$$

As before $N_{b.min.Rd}$ gives a conservative answer

$$k_z = 1,5 \text{ and } \eta = \frac{\gamma_{M0}}{\gamma_{M1}} \text{ for Class 1,2 or 3 cross-sections}$$
$$= 1,0 \text{ for Class 4 cross-sections}$$

Additional moments must be added for Class 4 cross-sections.

(c)     Axial compression and biaxial bending

$$\left[ \frac{N_{Sd}}{N_{b.min.Rd}} + \frac{k_y M_{y.Sd}}{\eta M_{c.y.Rd}} + \frac{k_z M_{z.Sd}}{\eta M_{c.z.Rd}} \right] \quad \leq \quad 1,0$$

For members which are susceptible to lateral torsional buckling,

$$\left[ \frac{N_{Sd}}{N_{b.z.Rd}} + \frac{k_{LT} M_{y.Sd}}{M_{b.Rd}} + \frac{k_z M_{z.Sd}}{\eta M_{c.z.Rd}} \right] \leq \quad 1,0$$

$k_y$, $k_z$ and $\eta$ are as defined before,

$$k_{LT} = 1 - \frac{\mu_{LT} N_{Sd}}{N_{b.z.Rd} \gamma_{M1}} \leq 1,0 \text{the value of } \mu \text{ is defined in C-EC3 but can be taken}$$

conservatively as 1,0 therefore $\quad k_{LT} = 1 - \dfrac{N_{Sd}}{N_{b.z.Rd} \gamma_{M1}}$

## 7.10   Example 7.5     Multi-storey column in simple construction

A multi-storey column in simple construction supports four beams at first floor level as shown in Figure 7.8. Check the suitability of a 203 × 203 × 60 UC given the factored design loads indicated.

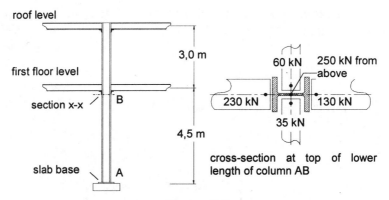

**Figure 7.8**

**Solution:**

Section properties: $203 \times 203 \times 60$ UC

| | | | |
|---|---|---|---|
| $h = 209{,}6$ mm | $b = 205{,}2$ mm | $t_w = 9{,}3$ mm | $t_f = 14{,}2$ mm |
| $c/t_f = 7{,}23$ | $d/t_w = 17{,}3$ | $d = 160{,}8$ mm | $i_{zz} = 5{,}19$ mm |
| $i_{yy} = 9{,}0$ cm | $i_{zz} = 5{,}19$ cm | $i_{Lt} = 5{,}53$ cm | $a_{LT} = 64{,}5$ cm |
| $a_{LT}/i_{LT} = 11{,}7$ | $A = 76{,}0$ cm$^2$ | | |

*Table 3.1*    Material strength              Grade *S275* steel              $t < 40$ mm

$$f_y = 275 \text{ N/mm}^2 \qquad f_u = 430 \text{ N/mm}^2$$

*Clause 5.1(1)*               $\gamma_{M0} = 1{,}05$               $\gamma_{M1} = 1{,}05$

*Table 5.6(a)*    Section classification

The outstand compression flange of the section is subject to compression only:

$$c/t_f = 7{,}23 < 9{,}2$$

**Flange is Class 1**

The web of the section is subject to both compression and bending and should be classified according to the criteria given in *Clause 5.3.2*.

*Clause 5.3.2.1*  For symmetric I -sections

$$\sigma_w = \frac{N_{Sd}}{dt_w}$$

At section x-x on column length AB

$$N_{sd} = (230 + 130 + 60 + 35 + 250) = 705 \text{ kN}$$

$$\sigma_w = \frac{N_{Sd}}{dt_w} = \frac{705 \times 10^3}{160{,}8 \times 9{,}3} = +471 \text{ N/mm}^2$$

*Table 5.8*    Limiting values of $\sigma_w$    (see  Extract from *Table 5.8*)

**Web is Class 1**

An alternative conservative approach is to classify the web as subject to compression only

i.e.    *Table 5.6(a)*    $d/t_w = 17,3 < 30,5$    **Web is Class 1**

**Section is Class 1**

---

*Clause 5.3.2.1*    For Class 1 and Class 2 webs which are subject to bending and tension or compression:

(1)    For symmetric I-sections the following equation should be satisfied:

$$\sigma_w = \frac{N_{Sd}}{dt_w} \leq \text{ } \textit{Table 5.8 } \text{ value}$$

where:

$\sigma_w$    is the mean web stress, taken as +ve for compression and −ve for tension

(2)    For rectangular hollow sections the following equation should be satisfied:

$$\sigma_w = \frac{N_{Sd}}{2dt_w} + \frac{|M_{z.Sd}|}{dt(b - t_w)} \leq \text{ Table 5.8 value}$$

where $\sigma_w$ as in (1).

*Clause 5.3.2.2*    For Class 3 webs which are subject to bending and tension or compression:

(1)    For symmetric I-sections the following equation should be satisfied:

$$\sigma_a = \frac{N_{Sd}}{A} \leq \text{ } \textit{Table 5.9 } \text{ value}$$

where:

$\sigma_a$    is the mean longitudinal web stress, taken as +ve for compression and −ve for tension

$A$    is the gross cross-sectional area.

(2)    For rectangular hollow sections the following equation should be satisfied:

$$\sigma_a = \frac{N_{Sd}}{A} + \frac{(b - t_W)|M_{z.Sd}|}{bW_{el.z}} \leq \textit{Table 5.9 value}$$

where $\sigma_a$ as in (2)

The section classification for Class 4 cross-sections is based on the '*effective*' width of the compression elements and is dealt with in *Clause 5.3.4*.

---

| Table 5.8 Limiting values of $\sigma_w$ (N/mm²) for Class 1 and 2 cross-sections [compression positive : tension negative] | | | |
|---|---|---|---|
| | Class 1 | | Class 2 |
| $d/t_w$ | Fe *430* (S275) ($t \leq$ 40 mm) | Fe *510* (S355) ($t \leq$ 40 mm) | Fe *430* (275) ($t \leq$ 40 mm) |
| 26 | * | * | * |
| 28 | * | 312 | * |
| 30 | * | 273 | * |
| 32 | 239 | 238 | * |
| 34 | 212 | 207 | * |
| - | - | - | - |
| - | - | - | - |
| * For Grade 275 steel with $d/t_w$ = 17,3 < 26     $\sigma_w$ is always satisfactory | | | |

Extract from Table 5.8 *C-EC3 Concise Eurocode...*(The Steel Construction Institute)

At section x-x the net moments are as follows:

Net moment about the y-y axis $= \left[ \dfrac{(230-130)}{10^3} \left( \dfrac{209,6}{2} + 100 \right) \right]$     = 20,48 kNm

Net moment about the z-z axis $= \left[ \dfrac{(60-35)}{10^3} \left( \dfrac{9,3}{2} + 100 \right) \right]$     = 2,62 kNm

*Appendix F, Clauses F.4(1)* and *(2)* of C-EC3 permits the net moment at any level of a multi-storey column in simple construction to be divided between the column length above and below that level in proportion to the stiffness coefficients I/L of each length. This is the same as *Clause 4.7.7* in BS 5950: Part 1 with the exception that the approximation using 50% of the moment when $\dfrac{(I/L)_{lower}}{(I/L)_{upper}} \leq 1.5$ is not included.

Consider the lower length of column at section x-x

$\qquad (I/L)_{lower} = (I/4,5) \qquad (I/L)_{upper} = (I/3,0)$

$\therefore$ stiffness coefficient for lower length $= \dfrac{(I/4,5)}{(I/4,5)+(I/3,0)} = 0,4$

design moment about y-y axis   $M_{y.Sd} = 0,4 \times 20,48 = 8,19$ kNm design moment about z-z axis $M_{z.Sd} = 0,4 \times 2,62 = 1,05$ kNm

design axial load $N_{S.d} = 705$ kN

*Clause 5.6.1.2* Resistance of Class 1 and Class 2 cross-sections with low shear
A conservative check can be made using *Clause 5.6.1.2(4)*, i.e.

$$\frac{N_{Sd}}{Af_y / \gamma_{M0}} + \frac{M_{y.Sd}}{M_{pl.y.Rd}} + \frac{M_{z.Sd}}{M_{pl.z.Rd}} \le 1,0$$

$$Af_y/\gamma_{M0} = (7600 \times 275)/(1,05 \times 10^3) = 1990 \text{ kN}$$

*Table 5.27*
$$M_{pl.y.Rd} = \frac{W_{pl.y.Rd} f_y}{\gamma_{M0}} = \frac{654 \times 10^3 \times 275}{1,05 \times 10^6} = 171,3 \text{ kNm}$$

$$M_{pl.z.Rd} = \frac{W_{pl.z.Rd} f_y}{\gamma_{M0}} = \frac{303 \times 10^3 \times 275}{1,05 \times 10^6} = 79,4 \text{ kNm}$$

*Clause 5.6.3.2(1)*
$$\eta = \frac{\gamma_{M0}}{\gamma_{M1}} = 1,0$$

$$\frac{N_{Sd}}{Af_y / \gamma_{M0}} + \frac{M_{y.Sd}}{M_{pl.y.Rd}} + \frac{M_{z.Sd}}{M_{pl.z.Rd}} = \frac{705}{1990} + \frac{8,19}{171,3} + \frac{1,05}{79,4} = 0,42 < 1,0$$

**Section is adequate in compression**

If this check had failed then a more exact approach could have been used using the expression in *Clause 5.6.1.2(3)*.

*Clause 5.6.3.4* Buckling resistance moment
(1) All members subject to axial compression and biaxial bending must satisfy

$$\frac{N_{Sd}}{N_{b.min.Rd}} + \frac{M_{y.Sd}}{\eta M_{c.y.Rd}} + \frac{M_{z.Sd}}{\eta M_{c.z.Rd}} \le 1,0$$

*Clause 5.4.3.2(1)*
$$N_{b.min.Rd} = \frac{\beta_A f_c A}{\gamma_{M1}}, \qquad \beta = 1,0 \quad \text{for Class 1 sections}$$

*Table 5,12*
Effective buckling length about the y-y axis = 4500 mm
Effective buckling length about the z-z axis = 4500 mm

*Clause 5.4.3.2(2)*
$$\lambda_{yy} = \frac{l_{yy}}{i_{yy}} = \frac{4500}{90} = 50$$

$$\lambda_{zz} = \frac{l_{zz}}{i_{zz}} = \frac{4500}{51,9} = 86,7$$

*Clause 5.4.3.2(5)*              $\lambda_{yy} < 180$          $\lambda_{zz} < 180$

*Table 5.13*      $h/b \leq 1,2$
for y-y axis use buckling *curve b*
for z-z axis use buckling *curve c*

*Table 5.14(a)*   for y-y axis  $\lambda\sqrt{\beta_A} = 50$       $f_c = 233 \text{ N/mm}^2$

for z-z axis  $\lambda\sqrt{\beta_A} = 86,7$     $f_c = 148 \text{ N/mm}^2$

**Critical value of $f_c = 148 \text{ N/mm}^2$**

$$N_{b.min.Rd} = \frac{\beta_A f_c A}{\gamma_{MI}}, = \frac{1,0 \times 148 \times 7600}{1,05 \times 10^3} = 1071 \text{ kN}$$

*Clause 5.6.3.2(1)*       Assume $k_y = 1,5$   (conservative value)
*Clause 5.6.3.31)*        Assume $k_z = 1,5$   (conservative value)

*Clause 5.5.2.2(1)*        $M_{c \cdot y.Rd} = \dfrac{W_{pl.y.Rd} f_y}{\gamma_{M0}} = 171,3 \text{ kNm}$

$$M_{c.z.Rd} = \frac{W_{pl.z.Rd} f_y}{\gamma_{M0}} = 79,4 \text{ kNm}$$

*Clause 5.6.3.2(1)*                 $\eta = \dfrac{\gamma_{M0}}{\gamma_{MI}} = 1,0$

$$\frac{N_{Sd}}{N_{b.min.Rd}} + \frac{M_{y.Sd}}{\eta M_{c.y.Rd}} + \frac{M_{z.Sd}}{\eta M_{c.z.Rd}} = \frac{705}{1071} + \frac{1,0 \times 8,19}{171,3} + \frac{1,0 \times 1,05}{79,4} = 0,72 < 1,0$$

*Clause 5.5.5(2) and (3)*     If $\lambda_{LT}\sqrt{\beta_w} > 34,7$ then lateral torsional buckling must be checked

*Table 5.22*      The value of $[k/C_1]^{0,5}$ is given for end moment loading for the ratio

$$\psi = \frac{\text{smaller end moment}}{\text{larger end moment}} = \frac{0}{8,19} = 0 \text{ and } k = 1,0$$

$$[k/C_1]^{0,5} = 0,73 \text{ Assume} \quad \beta_w = 1,0$$

*Clause 5.5.5(9)*

$$\lambda_{LT} = \frac{[k/C_1]^{0,5} L/i_{LT}}{\left[1 + \dfrac{(L/a_{LT})^2}{25,66}\right]^{0,25}} = \frac{0,73 \times \left(\dfrac{4500}{55,3}\right)}{\left[1 + \dfrac{(450/645)^2}{25,66}\right]^{0,25}} = 45,7$$

$\lambda_{LT}\sqrt{\beta_w} > 34,7$  ∴ lateral torsional buckling should be checked

*Clause 5.6.3.4(2)*     For members susceptible to lateral torsional buckling

$$\frac{N_{Sd}}{N_{b.z.Rd}} + \frac{k_{LT} M_{y.Sd}}{M_{b.y.Rd}} + \frac{k_z M_{z.Sd}}{M_{c.z.Rd}} \leq 1,0$$

*Table 5,18*     $\lambda_{LT}\sqrt{\beta_w} = 45,7,$     and     $f_y = 275 \text{ N/mm}^2$

$$f_b \approx 235 \text{ N/mm}^2$$

*Clause 5.5.5(7)*     $M_{b.Rd} = \dfrac{\beta_w f_b W_{pl.y}}{\gamma_{M1}} = \dfrac{1,0 \times 235 \times 567 \times 10^3}{1,05 \times 10^6} = 126,9 \text{ kNm}$

Assume $k_{LT} = 1,0$

$$\frac{N_{Sd}}{N_{b.z.Rd}} + \frac{k_{LT} M_{y.Sd}}{M_{b.y.Rd}} + \frac{k_z M_{z.Sd}}{M_{c.z.Rd}} = \frac{705}{1071} + \frac{(1,0 \times 8,19)}{126,9} + \frac{(1,5 \times 1,05)}{(1,0 \times 79,4)} = 0,74 < 1,0$$

**Section is adequate**

# Bibliography

1.  **BS 5950: Structural use of steelwork in building**
    Part 1: Code of practice for design of simple and continuous construction, BSI, 1990

2.  **BS 6399: Loading for buildings**
    Part 1:  Code of practice for dead and imposed loads (1984)
    Part 2:  Code of practice for wind loads (1995)
    Part 3:  Code of practice for imposed roof loads (1988)
    Code of practice for imposed roof loads – Draft for public comment – 1995
    BSI, 1984, 1988, 1995

3.  **Extracts From British Standards for Students of Structural Design**
    BSI, 1988

4.  **BS 648: Schedule of weights of building materials**
    BSI, 1964

5.  **CP 3: Chapter V  Loading**
    Part 2:  Wind Loads (1972) – Superseded by BS 6399: Part 2, BSI, 1972

6.  **Eurocode 3: Design of steel structures  DD ENV  1993 - 1 - 1**
    Part 1.1: General rules and rules for buildings, 1992

7.  **Baddoo N.R., Morrow A.W. and Taylor J.C.**
    *C-EC3 – Concise Eurocode 3 for the design of steel buildings in the United Kingdom*
    ,    The Steel Construction Institute, Publication Number 116, 1993

8.  **Taylor J.C., Baddoo N.R., Morrow A.W. and Gibbons C.**
    *Steelwork design guide to Eurocode 3: Part 1.1 – Introducing Eurocode 3.*
    *A comparison of EC3: Part 1.1 with BS 5950: Part 1*
    The Steel Construction Institute, Publication Number 114,1993

9.  **Narayanan R., Lawless V., Naji F.J. and Taylor J.C.**
    *Introduction to Concise Eurocode 3 (C - EC3) – with Worked Examples*
    The Steel Construction Institute, Publication Number 115, 1993

10. **Introduction to Steelwork Design to BS 5950: Part 1**
    The Steel Construction Institute, Publication Number 069, 1988

11. **Steelwork Design Guide to BS 5950: Part 1: 1990**
    Volume 1 Section Properties, Member Capacities
    The Steel Construction Institute, Publication Number 202, 1990

12. **Steelwork Design Guide to BS 5950: Part 1: 1990**
Volume 2, Worked Examples (Revised Edition)
The Steel Construction Institute, Publication Number 002, 1990

13. **Supplement A: Revised I and H Sections to Steelwork Design Guide to BS 5950: Part 1: 1990 – Volume 1 Section Properties Member Capacities (3rd Edition)**
The Steel Construction Institute, Publication Number 144, 1994

14. **Joints in Simple Construction**
Volume 1: Design Methods (1991)
Volume 2: Practical Applications (1992)
The Steel Construction Institute, 1991, 1992

15. **Joints in Steel Construction – Moment Connections**
The Steel Construction Institute, Publication Number 207/95, 1995

16. **Design for Manufacture Guidelines**
The Steel Construction Institute, Publication Number 150, 1995

17. **Steel Designers' Manual, Fifth Edition**
Blackwell Science (UK), 1994

18. **Handbook of Structural Steelwork**
The Steel Construction Institute, Publication Number 21/90, 1990

19. **Hayward, Alan and Weare, Frank**
*Steel Detailers' Manual*, Blackwell Science (UK), 1992

20. **Baird, J.A. and Ozelton, E.C.**
*Timber Designers' Manual*, 2nd Edition, Blackwell Science (UK), 1984

# Index